Practical
DRAINAGE
for
GOLF,
SPORTSTURF,
and
HORTICULTURE

Keith McIntyre BSc. Hons. MSc.
and
Bent Jakobsen PhD

Ann Arbor Press
Chelsea, Michigan

Library of Congress Cataloging-in-Publication Data

McIntyre, Keith
    Practical drainage for golf, sportsturf, and horticulture / Keith McIntyre
    and Bent Jakobsen.
        p.   cm.
    Includes bibliographical references.
    ISBN 1-57504-139-1
    1. Soil percolation. 2. Turfgrasses. 3. Synthetic sporting surfaces. 4.
    Drainage. I. Jakobsen, Bent, 1934– II. Title.
    S594 .M35 2000
    631.4'32—dc21   99-053388

ISBN 1-57504-139-1

PRINTED IN THE UNITED STATES OF AMERICA
10 9 8 7 6 5 4 3 2

Acknowledgments

The authors wish to acknowledge the following for their assistance in the preparation of this book.

Fiona Edge from Design Edge and Peter Child for illustrations; Rod Usback, Julia Rudolph, Charles Ollerenshaw, and Sue Johnston for editing and helpful advice.

Special thanks to my wife, Gale McIntyre, for editing, transport, and, most importantly, extreme tolerance during the gestation of this book.

The Authors

Keith McIntyre has a BSc Hons from the University of New England and a MSc from the Australian National University. He has 34 years of experience in horticulture and sportsturf. He worked for the Australian National Botanic Gardens for 8 years, and then moved to City Park's Technical Services Unit in Canberra, a specialist group working in soils, drainage, irrigation, pest management, tree management, and urban lake management. Keith worked there for 19 years and managed the Unit for 5 years.

This Unit built up a reputation for excellence in cool season turf management and irrigation, but its most important contribution was in the field of sportsground construction, soils, and drainage.

Keith left the ACT Government in 1995, and set up his own consultancy—Horticultural Engineering Consultancy in Canberra. His current main area of expertise is in sportsturf and the associated areas of design, profile design, drainage, and soil selection. He has recently been involved in seven of the new facilities being constructed for the 2000 Olympics.

Dr. Bent Jakobsen has a BAg and a PhD in Soil Science from the Royal Veterinary and Agricultural University in Copenhagen. He has spent much of his working life working on compaction in agricultural soils, and on particle movement within soils. He worked for CSIRO in Adelaide at the Waite Agricultural Research Institute on agricultural soil compaction and with the CSIRO

Division of Forest Research on soil compaction problems associated with logging operations.

Bent spent six years at Canberra's Technical Services Unit working with Keith on soil and drainage problems. He has been responsible for developing several simple, but very effective, laboratory testing techniques to determine the compacted hydraulic conductivity of soils, and has related these tests to the real world of sportsturf use.

His in-depth knowledge of soil physics and his ability to apply this science to the development of sportsturf profiles has made a unique contribution to the industry.

The team of Keith McIntyre and Bent Jakobsen has been responsible for designing the profiles of many of Australia's elite sports stadiums over the past ten years. These have all been sand-based perched water table constructions. These include the Suncorp and QEII stadiums in Brisbane; Stockland Stadium in Townsville; the Australian Rules facility in Cairns; the AIS soccer and athletics facilities, the National Baseball Arena, the roof of Parliament House, and the Ainslie Football Club and Tuggeranong football arenas in Canberra. Seven other sand-based facilities have Reflex® mesh elements reinforcement included in the profile. These are the Melbourne Cricket Ground and Moonee Valley racecourse in Melbourne; Bruce Stadium in Canberra; and Parramatta Stadium, Belmore Sportsground, the Showground, and the Olympic Stadium in Sydney.

Their methodology has also proved successful for a large number of golf and bowling greens all around Australia.

They have been involved as well in the profile design and the soil and drainage specification for more than 50 ha of traditional sportsgrounds; and Wagga, Hawkesbury, Goulburn, and Bega racecourses.

In urban landscape and horticulture they have developed specifications for soils for lawns and shrub and flower beds, as well as the production of compost for Canberra's Floriade. The strict adherence by the engineering profession to these specifications has resulted in a marked improvement in the quality of facilities, both in drainage and usability.

Introduction

Drainage as it relates to horticulture and sportsturf is a subject that is widely misunderstood and badly taught at many levels of education. This is not surprising, because the principles on which it is based are derived from complex aspects of soil physics.

Historically people have tried to make learning about drainage easy by putting it in a simple form that can be understood by students who have little or no knowledge or understanding of physics. In many cases the teachers themselves have limited knowledge of physics.

In the process of these simplifications, incorrect assumptions have often been made. The most frequent of these is that water will move sideways in soil at a reasonable rate. One may sensibly ask "Well, why wouldn't it?" It does seem logical, but in fact it is *wrong.* Water moves sideways in soil extremely slowly, and for all intents and purposes the significance of this movement can be ignored in most situations.

If this initial incorrect assumption has been made—*that water will indeed move sideways in soil at a reasonable rate*—the decisions that follow based on this assumption often have quite serious and expensive consequences.

We see engineers, landscape architects, and other professionals installing subsoil drains in certain locations, making the assumption that water will "quite logically" flow into them through the soil. In a large number of cases these drains will never drain the area they are supposed to, simply because the first assumption that "water flowed sideways in the soil at a reasonable rate" was itself *wrong.* Sadly, some textbooks used in lower level practical education teach or infer that these subsoil drains will drain certain areas, and they are wrong.

On the basis of this incorrect assumption, we see millions of dollars being spent all around the world by people installing subsoil drains in playing fields, golf courses, racecourses, garden beds, and various other places. Unfortunately, these will never drain anything effectively other than the area directly above the drainpipe itself, and perhaps with time a total width of a few feet on either side of the drain if the soil profile is deep enough.

The concept of a *perched water table,* which occurs when a fine soil, or sand, is placed over the top of a coarser material (usually sand or gravel), is a very difficult one to grasp. This concept seems to defy logic, as any reasonable and logical-thinking person would assume that water will flow from a fine soil into a layer of coarse sand or gravel below it. They are *wrong.* When a fine soil or sand is placed over a coarser sand or gravel and there is a sharp interface between the two, *no* water will flow from the fine soil into the coarse material below it until a saturated layer is built up at the interface. Eventually water is forced through into the layer below, but this saturated layer or perched water table will always remain. If the topsoil is too fine, the depth of this saturated zone, the perched water table, may extend to the surface of the topsoil and the whole profile will remain permanently saturated.

As this concept defies apparent logic or **common sense,*** it has led to a large number of disastrous consequences in horticulture and sportsturf, which have often been very costly. The wrong combination of soils in this situation has led to playing fields becoming quagmires, shrub beds and flower beds becoming "water beds," and planter boxes and nursery pots not draining.

The concept that *"soils that look alike will behave alike"* is also wrong, and often has serious consequences. How often do we hear or see references to a sandy loam, or a fine or coarse sand, in general horticultural and sportsturf jargon?

While to the eye two soils may appear to be the same, they may behave quite differently because of small differences in their particle size distribution, or their clay or silt content. One result may be good and the other can be a disaster.

There is also a perception that if a soil is sandy it will drain well. In certain situations this has proved to be very costly when the wrong "sandy soil" was used to build a playing field or golf green. It is worth remembering that *"sands ain't sands."*

The behaviour of a soil can be predicted by carrying out laboratory tests such as particle size analyses and hydraulic conductivity at various compaction levels. These tests can give a very good idea of

---

* Common sense does not exist—it is a decision based on prior knowledge.

how a soil will behave in certain situations in the field. This professional approach should be essential to ensure that a new facility will perform in the way that it was intended. No good road engineer would build a road these days without having all the components tested in the laboratory. They would also have no compunction about rejecting any materials that did not meet strict specifications and may lead to the failure of their road in the future.

The landscape construction and sportsturf industries will have to attain the same professional standards and attitudes as the road engineers if they are to consistently produce a better product and eliminate the very frequent and costly failures that currently occur in our industry.

Fundamental to this happening is a much better understanding of how water behaves in soil, and the concepts of soil structure and drainage as a whole.

One can see from the three examples already touched on in this introduction that the industry is dealing with a highly sophisticated subject that is quite difficult to simplify without making some serious mistakes. There are, however, some good simple tests that can be carried out, and some rules that can be made. If these are strictly adhered to they will greatly improve the decision-making for the practitioner, as well as the professional designer. These measures will also rule out some of the serious mistakes that have been made in the past.

This book is intended to give teachers a lead and to reduce the amount of misleading information that has been taught in the past. The authors hasten to add that this is usually not the teachers' fault because the textbooks are often incorrect, and the teachers may have sometimes been given incorrect and misleading information in their own education.

We hope this book will be the beginning of a reform in the understanding and teaching of soils and drainage in the fields of landscape architecture, sportsturf construction and management, and horticulture in general.

Also, when civil engineers come to design and build sportsfields and other horticultural facilities, they will see that there are quite different approaches required in drainage, compared to those used in building roads and other civil works. The one major difference is that in horticulture we want the water to get into the subgrade and be drained downward, whereas in road engineering the base is compacted to prevent the entry of water.

In horticulture and sportsturf the aim is to grow a crop, and therefore we need the soil to accept water. It follows that when this topsoil becomes waterlogged it has to drain downward through a largely uncompacted subgrade.

The authors hope that this book may help the engineering profession to produce sporting and horticultural facilities that are not built like roads.

Table of Contents

Practical
DRAINAGE
for
GOLF,
SPORTSTURF,
and
HORTICULTURE

CHAPTER 1

Soil Structure

It is not possible to assess the more complex issues of drainage without first understanding the basic properties of soil. Emphasis should be given to the physical properties of soil and to how these provide the framework for grass and other plant growth. It also gives us an insight into how particular soils may behave in a sportsturf or landscape situation, what happens when they are used in construction or played on under wet conditions, and how compaction will affect their ability to drain.

Particle Size Distribution

The behaviour of water in soil depends on the structure of that soil. This depends on the size of the soil particles, their arrangement, and the size and distribution of the voids, or pore spaces, within the soil.

Soil consists of primary particles of very widely differing sizes, as shown in Table 1.1. Some particles bond together and behave as much larger particles and form secondary soil structures. These groups of particles have a major impact on soil behaviour. Clay particles aggregate, and these aggregates occur in several different forms (Yong and Warkentin, 1975).

The most important forces holding particles and secondary soil structural units together at field water content are bonds due to nonclay material, either inorganic or organic, bonding to surfaces of more than one clay particle. Iron oxide, aluminium oxide, and carbonates are the most important of the inorganic bonding materials

Table 1.1. Different soil particles (and their diameters), and familiar objects with similar relative sizes.

Soil Particles	Diameter (mm)	Objects with Similar Relative Sizes
gravel	> 2	
very coarse sand	1–2	soccer ball
coarse sand	0.5–1	tennis ball
medium sand	0.25–0.5	golf ball
fine sand	0.1–0.25	play marble
very fine sand	0.05–0.1 ⎫	match head
silt	0.002–0.05 ⎬ **Fines**	sesame seeds
clay	< 0.002 ⎭	table salt

(Yong and Warkentin, 1975). Precipitates of calcium and magnesium carbonate can form between particles, bonding them together. Organic matter also bonds soil particles together to form larger particles. Organic matter can bond with clay particles and aggregates, as well as binding together larger particles, such as sand.

The stability of such aggregates is extremely variable and depends on the origin of the minerals that make up that soil's particles. When these materials aggregate they behave as larger particles, not as fine silt or clay particles. Thus a well structured (well aggregated) clay soil may drain much better than a soil that has far less clay, but possibly a higher silt content.

There is a difference in size between the primary soil particles of the order of 1000, which leads to an enormous difference in the way these particles will fit together and behave. The particles classified in Table 1.1 as *fines* are very important, as they have the ability to move within the soil when it is saturated.

The clay particles have a special significance, as they consist of minerals with special properties that attract cations, hence their importance in plant nutrition.

In the geomorphological processes (including soil formation) that created our current land forms, these differing soil properties played a very important role.

Particles that were carried down by rivers and then deposited on floodplains and in lakes often formed yellow or gray soils. Such soils have a very low stability and readily disperse when immersed in water. The particles break away from each other and are suspended in solution. Usually they have a high silt content.

On the other hand, many red or brown soils from hill slopes, and also some black clays, have a very high stability. This is because the clay particles have been formed into larger aggregates, which are very stable and do not disperse into primary particles when immersed in water. Such soils require intensive mechanical treatment to be dispersed, even for particle size analysis in a laboratory.

It appears that during erosion processes the soil particles are sorted out according to their stability. The most stable resist erosion, while the less stable end up in the water of streams, rivers, and lakes. Thus lake silt and floodplain soils are usually undesirable components in soils for sportsfields or for other horticultural purposes. This is because they are usually quite unstable and subject to a great deal of particle movement and compaction when they are worked or played on when wet.

Unfortunately, most of the soil used for the above-mentioned purposes comes from river deposits, causing drainage and compaction problems after construction.

Dispersion Test

A very simple test can be carried out to show if a soil is unstable and prone to dispersal. This is called the Emerson Test (Loveday, 1974; Standards Australia, 1997), which generates a number between 1 and 8 (the Emerson number). The least stable soil is rated 1, and the most stable soil is rated 8. The method used is quite simple. Take a small (5 to 10 mm) chunk of soil and carefully place it into a small glass or beaker of distilled or deionised water. If the water immediately becomes cloudy around the soil, then it is very unstable. The cloudiness may take up to an hour to form with more stable soils, and if the water remains clear, and the lump of soil remains intact, it is stable.

Soils that show instant instability will almost certainly compact badly in sportsturf and other horticultural situations, and should not be used. Unfortunately, even experts cannot tell you how a soil will perform just by looking at it. Proper laboratory tests must be conducted to determine how a soil will behave. (See Figure 1.1.)

Figure 1.1. Stable and unstable soils immersed in distilled water. The unstable soil disperses and makes the water cloudy, while the stable soil does not disperse.

Pore Spaces

Soil particles fit together in an imperfect way, leaving voids between the particles. Hence soil can be considered as a network of channels filled with air and water and bounded by solid surfaces. Its fundamental properties depend on the geometry of this interconnected network called the pore spaces (Russell, 1962).

In the formation of soil or its rearrangement during construction of facilities, the voids between the larger particles or aggregates are not completely filled by smaller particles. These voids are called pore spaces. The size of individual pores depends on the size of soil particles and on how densely the particles are packed together. Where the fines are all packed between the larger sand grains and aggregates, the total pore space is small and the sizes of the pores themselves are very small. This often occurs in unstable alluvial soils mentioned earlier. Such dense interpacking is undesirable for plant growth, as there is little room for air and for root growth between the particles.

If the clay and fines are assembled in aggregates there will be a system of large-sized pores between the aggregates (macropores), plus a system of smaller pores inside the aggregates (micropores). Such

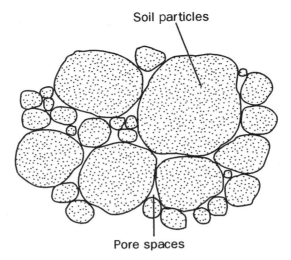

Soil particles

Pore spaces

Figure 1.2. Soil particles and pore spaces.

soils will drain well and are ideal for plant growth because there is much more air space and much more room for roots.

Whether one or the other of these arrangements of particles (soils) exists depends on the degree of soil compaction and on the stability of the soil aggregates. (See Figure 1.2.)

The pore spaces are very important in the behaviour of soils because they provide the pathways for the downward movement of water and the spaces in which roots grow. Pore spaces also provide the air space that is essential for good plant growth. Water is stored in the smaller pores, which are important in determining the water holding capacity of soils.

The number and size of these pores largely determines the drainage properties of soils.

CHAPTER **2**

Water Intake and Retention in Soils

This chapter deals separately with each of the following aspects:

- the entry of water into soils (***infiltration***)
- the rate at which water enters soils (***infiltration rate***)
- the movement of water through soil (***hydraulic conductivity***) and capillary movement
- the storage of, and depletion of water in soils (***saturation, field capacity,*** and ***wilting point***)

All these aspects are interrelated and are constantly changing in the soil. It is these changes, such as the slowing down of the entry rate of water into the soil, and the rate at which a saturated soil drains to reach field capacity, that actually constitute drainage.

To understand the behaviour of water in soils it is necessary to begin with the fundamental concept of how water moves in soils. Water movement in soils, and hence drainage, is influenced by the following forces:

- gravity
- surface tension of the water
- water adhesion to particles

Gravity

Gravity is the driving force of water movement in the soil by pulling water ***downward*** through it. As the depth of the profile increases, the height of the water column being pulled down by gravity increases, as does the weight of water. Understanding the effects of gravity in soils is *fundamental* to the understanding of how water moves in soil. Even when water is moving sideways in soil, except for capillary movement, it is being moved by gravity.

Surface Tension

Water molecules within a body of water are attracted to each other in all directions. Where water meets air a surface tension exists because all the water molecules are attracted to each other much more strongly than they are attracted to the adjacent air molecules (Marshall et al., 1996). (See Figure 2.1.)

This process *shrinks* the surface of the water, acting like a *drawstring* pulling the surface of a water drop together. This makes a water drop round as the surface tension shrinks it, and holds it together. The smaller the drop, the stronger the surface becomes, and the more difficult it is to break.

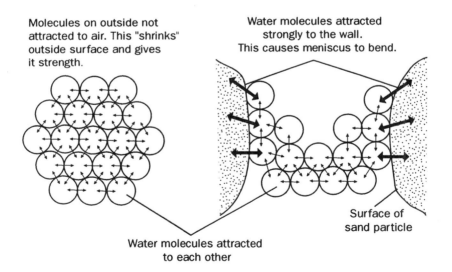

Figure 2.1. Water molecules being attracted to each other to form a drop and a meniscus. This figure also shows the meniscus being bent by the force of adhesion to the surface of particles.

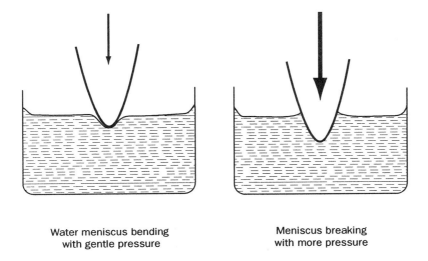

Water meniscus bending
with gentle pressure

Meniscus breaking
with more pressure

Figure 2.2. Water meniscus bending with gentle pressure, and breaking with too much pressure.

It also makes the surface of a water meniscus strong. Small particles, such as dust, can be observed sitting on the surface of water, or small insects can walk on the surface of the water without breaking the meniscus. The strength of the surface tension bonding the water molecules together forming the meniscus is sufficient to support the dust or the insect. The surface of water can be bent by gently touching it with an object, but if the force applied by the object is too strong the surface will break. (See Figure 2.2.)

Water Adhesion to Soil Particles

Water adheres to soil particles very strongly—much more strongly than to other water molecules.* The strong attraction of the water molecules to solid surfaces bends a meniscus where water is in contact with both air and solids. (See Figure 2.1.) The combined force of

---

* Note: In some instances, soil particles may become covered with a wax that makes them water-repellent. Then the attraction between water molecules is stronger than to the wax. This is the common cause of water repellence in sands.

Water held by surface tension on the particles
against the force of gravity

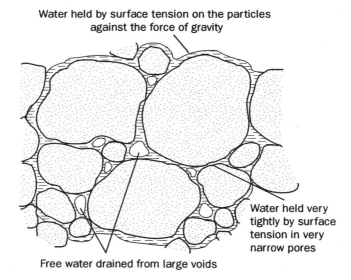

Water held very
tightly by surface
tension in very
narrow pores

Free water drained from large voids

Figure 2.3. Water held in small pores by the combined forces of surface tension and adhesion, and water drained out of the larger voids.

surface tension and adhesion can hold water back in small pores against the pull of gravity.

When soil, which is an aggregation of a range of particles, is formed, the particles arrange themselves in an imperfect way. As mentioned earlier, this leaves spaces between particles—the pores. When the soil is made up of predominantly large particles, such as a coarse sand, the pores will be large, but with soils where the particles are very small, such as a silt, the pores will be very much smaller. (See Figure 2.3.)

Water **_adheres_** to soil particles, and the gaps between the particles (pores) are spanned by menisci. These menisci are held together by surface tension, as explained previously. The closer the particles, i.e., the finer the pores, the more tightly the water is held, because there is a greater component of adhesive forces than surface tension holding the water in the pore. These adhesive forces are much stronger than surface tension.

The water in soil exists in two forms. One is a surface film around the particles held there by adhesion, while the other is the water held in the pore spaces by the surface tension of the menisci at the top and bottom of these pores.

When gravity is pulling water downward in the soil, these other two forces are acting against it and attempting to hold the water molecules near to the soil particles, or in the pore spaces. Gravity is stretching the menisci and bending them inward toward the surfaces of the particles. The more curvature there is on the menisci the stronger they become, which means that more force is required to remove further water from those pores.

Surface tension is the weakest of these forces and is the first to break under the pull of gravity. Hence, surface tension is the limiting factor in the amount of water a pore can hold.

The majority of the water in large pores is only held by surface tension, as it is too far away from the edges of the pores to be influenced by the adhesive forces. Hence it is easily pulled down by gravity, causing most of the water in these large pores to be drawn down quite quickly. The remaining water in these pores is attracted much more strongly to the edges of the particles by the stronger adhesive forces and requires greater force to remove it. As already explained, the water in the smallest of the pores will generally not be drawn down by gravity.

When a soil is saturated, all the pore spaces are filled with water and the menisci become flat surfaces. Gravity is then easily able to pull water downward; *it begins to drain.* (See Figure 2.4.)

The *rate* at which water moves downward through the soil is strongly affected by the pore size through which it has to move. This rate is approximately proportional to the square of the pore diameter; e.g., water will move at least 100 times faster through a USGA sand than a loam.

Thus as the water content of a soil decreases due to the pull of gravity, the rate of drainage decreases at an exponential rate, as the remaining water is held in the narrower pores.

A very important concept is noted here—once the soil is no longer saturated, water is prevented from moving downward through the soil at the rate of the saturated hydraulic conductivity. This is because even though there are large pores in the soil below the wetting front, water is prevented from filling them because it is being held in the smaller pores above by the forces of adhesion, and by the surface tension of the menisci at the ends of the smaller pores.

If the soil above it were saturated, water would rapidly enter and fill these large pores, and move downward faster.

When soil is saturated the menisci are
flat and the soil is full of water.

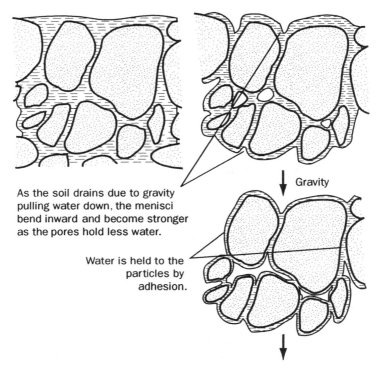

Gravity

As the soil drains due to gravity
pulling water down, the menisci
bend inward and become stronger
as the pores hold less water.

Water is held to the
particles by
adhesion.

Figure 2.4. At saturation the pores are full of water and the menisci are flat. Drainage begins as gravity bends the menisci against the forces of surface tension and adhesion. This continues until the menisci are all bent and water is only held in the small pores and gravity can no longer remove water.

Field Capacity

When the top part of a soil profile is saturated, either by rainfall or irrigation, all the pore spaces are filled with water. When the rain or irrigation ceases, this water is pulled down through the profile by the force of gravity. "Free water" is removed in this process.

Water is held in the smaller pores and around particles by adhesive forces and by the surface tension of the menisci. As the water is pulled downward by gravity from the larger pores, the remaining water in the smaller pores forms bridges (menisci) between neighbouring particles.

A point is reached when no more water can be drawn down by gravity, because it is being held in very narrow pores by the combined forces of adhesion onto the soil particles and surface tension of the water in the menisci. This state is called *field capacity.* (In reality, very small amounts of downward movement of water may continue, but this will be at rates of less than 1 mm per day, and are effectively taken to be zero.)

Therefore field capacity can be broadly defined as the moisture status of a soil which, having been saturated or nearly saturated, has freely drained to a point where gravity can no longer pull any more water downward from that part of the profile.

Continued water removal from the soil at this stage is by plants "pumping" the water out by transpiration. Plants cease extracting water at what is called the *wilting point,* at which point they may die. However, at this point it should be noted that there is still water in the soil, but it is held very strongly onto the particles and in the small pores by the adhesive forces. When the wilting point is reached, the conductivity of this water is 10,000 times smaller than when the soil is at field capacity.

For the top 4 in. (100 mm) of a soil profile, field capacity usually occurs after the soil has been saturated, then allowed to drain freely for about 24 hours (given that there is sufficient depth in the profile to accept the water from above). The amount of water remaining in the soil at field capacity will vary quite markedly, depending on the soil's particle size distribution. A loamy soil may have 25 % water remaining, a USGA sand may have about 5 to 7 %, and a gravel may have 2 % of water left—yet they have all drained to their own particular field capacity.

This only occurs if the soil profile contains a uniform texture and has a reasonable depth. If the base is of a finer material, such as clay, drainage may take much longer.

The amount of water retained by a soil at field capacity is also determined by the number of large and small pore spaces, remembering that the large pores will always drain very quickly. It is also worth noting that these large pore spaces are filled with air when the water drains out, and the amount of air space in a soil at field capacity is also very important for the growth of plant roots.

Plant roots require pores in excess of 0.1 mm in diameter to be able to easily force their way through the soil, as well as requiring sufficient oxygen for growth. If the soil is a dense silty one with a large number of very fine pores, not only will water move down through it very slowly, but it will not be very suitable for good root growth. These two characteristics of drainage rate and root growth are very strongly linked in soils.

Soils vary markedly in their pore space and sizes of pores, hence their behaviour in relation to water movement varies. For example, a heavy silt/clay soil has very small pores and water moves very slowly through it, whereas water moves very rapidly through a coarse sandy soil with very few fines and many large pores.

"Gravitational Water"

Gravitational water is that water in the soil that, after rain or irrigation, can be moved down through the profile by gravity (Adams and Gibbs, 1994). When field capacity has been reached, gravity will have removed all the gravitational water and an equilibrium will have been reached.

Soil Strength

As the content of the water in the soil changes, so too does the soil stability, and *strength.*

When soil is saturated it is quite unstable because the particles are not being held together by surface tension and the adhesive forces. As the water drains downward and is removed from the larger pores, the water in the smaller pores is held much more strongly by adhesion and the surface tension of the menisci. The steeper the curve on the meniscus of the water in the pore, the more tightly the water is held, and the more tightly adjacent soil particles are held to each other; i.e., the stronger or harder the soil becomes.

Thus, as the water is progressively removed, the soil becomes stronger, i.e., the soil particles become more tightly bonded together by the force of adhesion and surface tension. If the soil is dried out completely (the last remaining water has to be removed with heat) it becomes unstable again and becomes dust, with all particles becoming free from each other.

Conversely, as the water content of the soil is increased, the strength of the "bond" is reduced, and the soil becomes progressively softer. Mud can form when the soil is saturated, and when even more water is added it becomes a slurry. Once again, all soil strength and stability is lost.

There are other substances in the soil that bind or "glue" particles together, such as organic matter and the oxides of iron and aluminum. The main point here is that when soil is completely dry it can be easily broken up and its texture destroyed, breaking down into its basic particles. The same applies when it has too much water added, and it becomes saturated.

While other material, such as organic matter, may bind smaller soil particles together, the *water content* of a soil largely determines its stability and strength.

Suction

The forces that are required to pull water down through the soil, or those needed by plants to take up water against the forces of surface tension and adhesion, are commonly referred to by soil physicists as *suction.* Greater suction is required to get water out of a fine silt/clay soil than from a clean sandy soil. This is because fine silt/clay soils have smaller particles and pores than sands, hence they have a higher water holding capacity.

Saturation

The concept of saturation is very important in understanding a whole range of soil/water relationships in drainage and turf management. (See Figure 2.5.)

A soil, or part of its profile, becomes saturated* when all the pores have become filled with water. This can occur after rain or irrigation.

---

* Note: In the field, a "saturated" soil, or part of a profile may not have all its pores filled with water as there is always air trapped in some pores. However, for the purposes of discussion here, these are disregarded. It is also noted that much of the "saturated" zone in A, B, and C in Figure 2.5 is actually not saturated and is called the transmission zone. This is because the water in this zone is under some suction from the drier soil below, hence water is prevented from entering some of the larger pores (Yong and Warkentin, 1975). For the sake of a simple explanation of a very complex area of soil physics, this has also been disregarded.

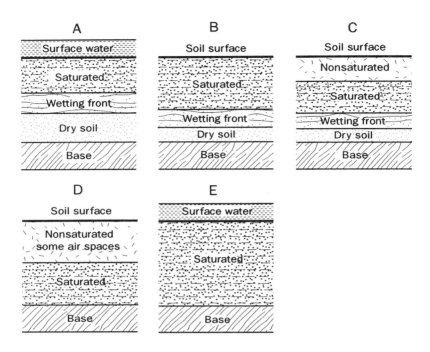

Figure 2.5. Saturation in soil is dynamic, and varies with time within a shallow soil profile.

Saturation may occur throughout the whole profile, or it may occur as a band at varying depths, as water moves down through the profile.

Saturation is a dynamic process and is always changing with time as water drains through the profile. From Figure 2.5 it can be seen that the location of a saturated zone is continually changing in a soil profile during and after rain. In A there is a saturated zone at the soil surface, and there is ponded water above the surface. As this surface water drains down into the profile, the saturated zone moves progressively down until in D it reaches the base. If there is sufficient rain after this occurs, as in E, the whole profile becomes saturated and water ponds on the surface again.

Infiltration and Infiltration Rate

Infiltration is the downward entry of water into the soil, and how this occurs, together with the rate at which it enters (*infiltra-*

tion rate), is vital in the understanding of drainage and irrigation practices.

Infiltration Rate

When water reaches the soil surface, either as rainfall or irrigation, it enters it at a particular rate, which is called the infiltration rate. This rate varies, depending on two factors:

a. **Soil texture**—The rate will be slow in a heavy clay soil and much quicker in a very sandy soil because there are more large pores in the sandy soil.
b. **Moisture status of soil at the time rainfall or irrigation commences**—If the soil is very dry the water enters quickly. If the soil is moist it enters more slowly, and the rate of entry continues to slow down as saturation is approached.

Figure 2.6 illustrates how the infiltration rate of a soil changes with its moisture status. Curve A shows that when the soil is at the grass stress point, and rainfall or irrigation commences, water enters the soil very quickly because all of the large pores are empty and water enters them very quickly. Thereafter, as more and more of the smaller pores fill up, the rate of entry will gradually slow down until the top of the profile is saturated. The rate at which water then enters the soil becomes constant. This constant rate is called the *saturated hydraulic conductivity*, as shown by curve C.

Curve B illustrates how the initial rate of entry of water is lower than A, when the soil is at field capacity at the commencement of rain or irrigation. In this situation the saturated hydraulic conductivity is reached with the application of less water. This is because at field capacity, while most of the larger pores are empty, much of the remaining pore space in the soil is already filled with water before rain or irrigation commences.

The infiltration rates quoted in some textbooks and horticultural literature are usually the saturated hydraulic conductivity; i.e., the final rate of infiltration after the soil has been wetted to a considerable depth. Graphs showing infiltration rate as a function of time are usually produced from data obtained from deep uniform soils found in agriculture.

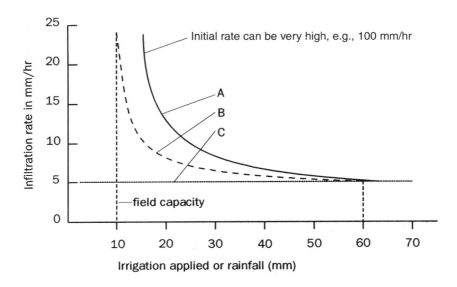

Figure 2.6. The infiltration rate of a silt/loam soil varies with three different moisture status levels at the time of commencement of rainfall or irrigation. A) the soil is at the grass stress point; B) the soil is at field capacity; and C) the soil is saturated.

These infiltration rate figures can only be considered as a minimum, particularly when considering how quickly a soil will accept irrigation when it is not saturated, and more particularly when it is below field capacity. They are inappropriate for calculation of infiltration rates for sportsturf irrigation.

In all turf situations irrigation should occur between the grass stress point (the point where grass is showing visible stress), and field capacity, i.e., between curves A and B. The rate at which soil accepts water when its soil moisture status is in this range is always much higher than the saturated hydraulic conductivity for that soil.

In practical terms the soil will accept much more water during the first cycle of irrigation than was previously assumed and recorded in some turf irrigation literature (McIntyre and Hughes, 1988). In the past some students were taught that where the estimated infiltration rate (usually the saturated hydraulic conductivity) was, say 0.2 in. (5 mm) per hour, irrigation should only be applied at about that rate, when in fact it could have been much higher.

For example, using this principle on the soil in Figure 2.6, if the precipitation rate of the irrigation system were 0.4 in. (10 mm)/hr, and if the infiltration rate (the saturated hydraulic conductivity) of the soil were 0.2 in. (5 mm)/hr, the first watering cycle should never have exceeded 30 minutes. From Figure 2.6, this is clearly wrong, because the rates from curves A and B show that soil probably could have absorbed water from a 45- to 50-minute cycle.

With sportsturf the situation is almost always different from that in agriculture, because the sportsturf profile is usually much shallower. Topsoil, ranging in depth from 3 to 8 in. (75 to 200 mm), is laid over a clay subsoil base, which is quite often compacted. In these conditions grass root zones rarely exceed 6 in. (150 mm) and on shallow topsoil profiles are as shallow as 1 in. (25 mm). This shallow topsoil and clay base combination has two effects:

First, the topsoil will take up water readily, and when it becomes saturated, further water entry into the soil will only be at the drainage rate of the compacted clay subsoil base, which could be as low as 0.04 to 0.2 in. (1 to 5 mm) per day. Second, the topsoil will become saturated to its entire depth by as little as 1 in. (25 mm) of rain, and certainly during a couple of days of rain.

Under these conditions the rate at which the topsoil drains will be determined by the rate at which the subsoil can remove it, i.e., very slowly, as this is the hydraulic conductivity of a much finer material than the topsoil. (See Table 2.1.)

Water initially enters the soil at a much faster rate than the saturated hydraulic conductivity. It is often possible for a soil to accept water initially at, say 1 in. (25 mm) per hour and have a saturated hydraulic conductivity of only 0.2 in. (5 mm) per hour. As shown earlier, this is particularly the case when the top part of the soil is dry at the time of the rain or irrigation. Under these conditions all the large pores will be empty, which is always the case if the soil is at field capacity or drier.

The rate at which initial water enters the soil is very fast, because the water is filling these large pores as well as the smaller ones. As soon as these large pores are full, gravity begins to pull water downward, and the rate at which this downward movement occurs will determine how much pore space is available for more rain or irrigation at the top of the soil profile. If the rate at which the rain is falling is faster than gravity can pull it further down the profile, the top of

Table 2.1. Shows how infiltration rate in in./hr (and mm/hr) varies between wilting point and saturation in four different soils from sportsfields in Canberra, Australia. When irrigation commenced the soil moisture content was at one of the following: permanent wilting point; the point where grass showed visible stress; field capacity; or the top of the profile was saturated.

Soil Type	Soil Moisture Content at Time of Irrigation	Permanent Wilting Point	Visible Grass Stress	Field Capacity	Top of Profile Saturated
sandy loam	in./hr	4.8	4.5	4.2	2.4
	mm/hr	122	114	107	60
loam	in./hr	1.4	1.3	1.2	0.8
	mm/hr	35	33	31	20
silt/loam	in./hr	0.7	0.6	0.5	0.2
	mm/hr	17	15	12.5	5
silt/clay	in./hr	0.3	0.21	0.14	0.04
	mm/hr	7	5.3	3.6	1

the soil will become saturated for a time. If water continues to reach the soil surface under these conditions, then *runoff* or *ponding* on the surface will occur.

The rate at which water enters the soil during rain or irrigation depends on the saturated hydraulic conductivity of the soil, gravity, and the suction applied by the soil, provided there is drier soil below the wetting front.

The moisture status of the soil at the time of rain or irrigation is very important in determining how quickly the infiltration rate will slow down to the saturated hydraulic conductivity.

The rate of water movement is proportional to the square of the diameter of the pores that are conducting water. With a range in particle sizes spanning three orders of magnitude, and a similar range in pore sizes, the rate of water movement may vary by six orders of magnitude.

The infiltration rate can be calculated using the following formula:

$$I = K (Z + S)/Z \text{ (mm/hr)}$$

where K = saturated hydraulic conductivity in mm/hr
     Z = depth of wet soil in mm
     S = soil water suction at wilting point in mm

The depth of wet soil (Z) is equal to the amount of water that has entered the soil divided by the water holding capacity of the soil. Often the depth of wet soil is three to six times the amount of water infiltrated: e.g., 0.25 in. (6.4 mm) of rain may wet to a soil depth 0.75 to 1.25 in. (19 to 38 mm).

The effective suction at the wetting front S is usually small, in the order of 20 to 200 mm of suction (water column), similar to the height of the capillary fringe. (See Figure 2.9 and later explanation in Chapter 5.) When the depth of wet soil exceeds this value, gravity is the dominating force determining the infiltration rate, and eventually this will be equal to the value of K in the equation. If the topsoil is shallow, i.e., less than 8 in. (200 mm), the suction (S) is important. If the topsoil is deep, then gravity will be the dominant force.

It should be emphasised that the field value of saturated hydraulic conductivity (K in the above equation), is usually much smaller than the values measured in the laboratory. This is because it is very difficult to simulate field conditions and because trapped air bubbles influence what happens in the field. These are excluded in laboratory techniques. The field value is often about one-quarter to one-third of the laboratory value, and grass roots occupying the larger pores in the field can make this difference even greater.

Water only moves very short distances horizontally in the soil, and any horizontal movement is extremely slow. The horizontal redistribution of water from wet soil into neighbouring dry soil in a sportsground situation would rarely exceed 9 in. (230 mm). (See Chapter 4.) This is important when considering uneven water distribution caused by poorly designed irrigation systems—soil moisture will never be uniform.

Sportsfields and golf fairways are often constructed with a shallow layer of permeable topsoil over a nearly impermeable base. This situation causes a sudden decline in infiltration rate when the topsoil becomes saturated. Once this occurs the final infiltration rate may be only a few millimetres per day, as determined by the rate at which the water moves down through the base.

Water distribution in the topsoil changes with time and with the intensity and duration of rainfall. The topsoil has a much higher hydraulic conductivity than the base, and as a consequence it accepts water faster than it can be discharged through deep drainage. The profile can be filled up completely when rain continues for days.

Water Storage in Soils

The soil profile is an ingenious system that can both absorb water very quickly and also store it for long periods in a form available to plants, particularly when the evaporation rate is low.

A good agricultural soil profile may absorb heavy rain falling at a rate of 2.5 in. (64 mm) per hour for up to half an hour. A few hours after the rain has fallen, this water will be redistributed down through the profile as the large pores near the surface drain quickly, and the upper root zone will begin to be aerated.

If the rain continues, the downward water movement slows and the whole profile may become saturated, particularly in a sportsfield situation. Thus water is retained within the reach of the plant roots for prolonged periods. If the subsequent evapotranspiration rate is high, such water may be used by the plants through transpiration, rather than lost as drainage. However, if saturation lasts for too long, poor aeration will result and root systems will die.

The percentage of pore spaces filled with water and that filled with air at a particular suction will depend on the size and distribution of soil pores.

Water Retention Curve (Moisture Release Curve)

The relationship between soil water content and suction is expressed by the water retention curve, as shown in Figure 2.7. The suction can be expressed as the length of a hanging water column or as a negative pressure, where 1 kPa equals 102 mm water column. In lay terms this refers to the depth of the soil profile being measured.

There are several points on the water retention curve that are of special interest. These are:

> **Air Entry Point.** At the air entry point the largest pores begin to drain. This occurs at suctions of a few millimetres and up to a few hundred millimetres, depending on the fineness of the soil. This point also shows the height of the *capillary fringe* (defined below). Below this point the soil can be

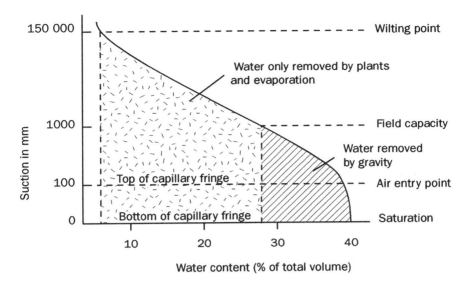

Figure 2.7. Water retention curve for a loamy soil. Water in the soil between field capacity and saturation is removed by gravity, but water in the soil between field capacity and wilting point can only be removed by plants and evaporation.

considered as saturated, i.e., all the pore spaces are filled with water. Above it the water has drained out of some of the larger pores and the soil has begun to be aerated again, i.e., these larger pores are being filled with air.

Field Capacity. At field capacity only the smaller pores contain water, and water movement under the influence of gravity has ceased (or for the purists, has slowed to a point whereby for all practical purposes it is now insignificant). The soil water suction at field capacity is typically around 1,000 mm (–10 kPa). It is less for sand and more for clay soil. Gravity has removed all the "free water" down through the soil, and the only water remaining is that which adheres to the particles and that held by surface tension (meniscal forces) in the small pores.

Wilting Point (WP). At the *wilting point* water is adsorbed so firmly in the soil, and its mobility is so low, that plant roots can no longer extract sufficient water to survive.

The wilting point is also defined by the water content at a suction of 150,000 mm (−1500 kPa).

The shape of the water retention curve depends mainly on the particle size distribution of the soil, and on the degree of soil compaction. The amount of water available to plants is that held at field capacity minus the amount of soil water at wilting point.

Water content at the wilting point is strongly related to the clay and organic matter (OM) content of the soil. A rough estimate of the wilting point can be made on this basis:

$$\textbf{WP (\%)} = 0.4 \times \text{Clay (\%)} + 0.7 \times \text{OM (\%)} \quad \text{(all by weight)}$$

When soil is compacted, the porosity is reduced and the amount of nonavailable water is increased. Light to moderate compaction mainly reduces the air volume of the soil, while heavy compaction also reduces the amount of water available for plants.

The amount of water available to plants is mainly dependent on the silt and very fine sand content of the soil, i.e., particles in the range of 0.002 to 0.1 mm.

The laboratory measurement of available water (the water retention curve) can be a useful guide in assessing the drainage characteristics of a soil. If too little water is released by the soil then it may not be suitable for growing grass; i.e., it may be droughty in the summer and become quickly waterlogged in the winter. This curve will be different for different soils, and will change with compaction.

Capillary Fringe

A *capillary fringe* (discussed further in Chapter 5) exists where there is a saturated zone at the bottom of the topsoil, immediately above a very slow-draining base. This situation usually occurs after prolonged rain, when the bottom of the profile becomes saturated—often a winter problem. (See Figures 2.5D, 2.5E, and 2.8.)

There is a zone at the top of this saturated section of the profile where water is held in the pores by surface tension and no water will move laterally out of this zone, e.g., into a drain or a hole.

If a hole is dug into the capillary fringe zone no water will enter this hole even though the surrounding soil is saturated. If, however, a

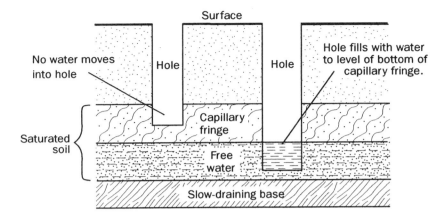

Figure 2.8. This shows no water moving sideways into a hole dug into the saturated capillary fringe, but water moving sideways into the hole that extends into the saturated "free water" zone.

hole is dug that goes below the capillary fringe, it will immediately begin to fill with water, but only to the level of the bottom of the capillary fringe. This is because there is sufficient "head" of water above it "pushing" the water out of the pores. This will continue, providing the head is greater than the surface tension of the menisci in the pores.

This principle is of great importance when understanding how water gets into drains. Its implications are discussed in Chapter 5.

Soils with a significant clay content may form aggregates and, if not compacted, such soils can have very large pores and a very small capillary fringe. However, when compacted their pore size decreases and the height of the capillary fringe will increase greatly, as it will now mainly be determined by the large number of fine pores. As the capillary fringe increases due to compaction very large changes can occur in the drainage characteristic of such soils.

Note

In this chapter the authors have endeavoured to define in lay terms many of the complex aspects of soil physics. These explanations may be at odds with some made by those in the soil physics profession, but a perspective should be kept.

Soil physicists will argue over the third decimal point on a wide range of complexities within the soil, and quite rightly so. These finer points, however, are quite often unnecessary in the understanding required for those in the horticultural and sportsturf profession, unless they already have a degree in soil physics.

As an example, when we define field capacity, the definition suffices for the horticulturist's pragmatic needs, while the purists will probably argue forever about its limits. Saturation is another case where the definitions in this book, and their use in other places, is adequate for the purposes intended.

Again, the authors have defined and used these often quite complex states as accurately as possible within the broad parameters of the "need to know" of our readership.

CHAPTER **3**

Compaction and Particle Movement

Compaction

The process of compaction is very important in understanding drainage in relation to sportsturf and general landscaping.

The physical properties of most natural soils are vulnerable to major changes when mechanical stress disrupts their aggregated structure (Adams and Gibbs, 1994). Almost all soils will compact; even soils that most people would call sand will compact. It is reasonable to assume that 99.9% of all soils used for the construction of sportsfields, school grounds, parks, and shrub beds will compact. This compaction is most likely to occur when the top of the soil profile is saturated or near saturation. Compaction can be explained simply as the rearrangement of the fine particles in the soil so that more of the larger pore spaces are filled with these "fines." This has several effects, as it increases the soil density, reduces the rate of movement of water through the soil, and reduces the amount of air space in the soil, which often restricts root growth. (See Figure 3.1.)

Traffic on soil applies a pressure to the soil particles, which brings them together and packs them more closely. These particles mostly stay in this closely packed state after release of the pressure. Vibrations and repeated application of pressure give the particles more opportunity for sideways movement, until they find a position where they fit more closely into a void. As this process continues, higher soil density results.

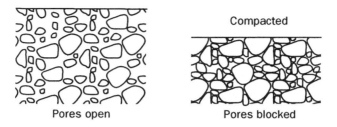

Pores open   Pores blocked

Figure 3.1. Soil density is increased and pore space reduced when a soil is compacted.

Vibration and shearing of the soil can also reverse the effects of compaction. If little or no pressure is applied to the soil at the time, vibrations may cause the particles to jump out of their dense packing. This is the process used by compaction breaking machines.

In shallow tillage where tines are pushing the soil aside and upward, a loosening of the soil results. In deeper tillage the weight of overlying soil applies a significant pressure, and in certain circumstances compaction rather than loosening may occur lower in the profile.

Major causes of compaction are often the result of the work processes during initial construction. Natural soils that may have evolved over millions of years may drain very well in their original state. These soils will behave quite differently after they have been dug up; placed on a truck; driven over a bumpy road; spread out thinly over a compacted clay base; run over with heavy machinery; and called a playing field, fairway, or even a shrub bed. It is no wonder that good natural soils often behave very poorly in relation to drainage when they are treated in this manner.

During all these processes soil structure and pore spaces are altered significantly, and soil aggregates may be smeared out and become a heap of primary particles. The rebuilding of aggregates is a slow process that may take years.

Biopores are channels in the soil left after decay of root systems, or made by burrowing soil fauna such as worms. Such pores are very efficient in the drainage and aeration of soil. They are destroyed by digging and compaction of soils; and as they are destroyed, the movement of water through the soil is reduced. This affects the "behaviour" of the fairway, playing field, or other facility.

The water content of a soil may largely determine its stability. When a soil is very dry it can often be easily broken up, degrading into individual particles, or dust. It can be very unstable in this state and is often blown away by the wind.

When soil is completely saturated, it becomes a slurry, is once again very unstable, and easily moved by water. In this unstable state it can be washed away into streams and rivers. Collectively, these effects of wind and water are called soil erosion.

The soil can have a wide range of moisture content, ranging from dust to slurry. The behaviour of a soil and its reaction to pressure will depend on the soil type and its moisture content. For example, when beach sand has a low moisture content it is very unstable and it is easy to bog a vehicle in it. Closer to the water the moisture content is higher, but not yet saturated, and it is quite stable and hard. When a vehicle is driven over it, it is well supported and the surface of the sand is barely marked. Go a little further into the water, and the sand becomes highly unstable again because it is completely saturated, and the vehicle will become bogged again.

Wet and saturated soil is much softer (has less strength) than dry soil and compacts under a much lighter pressure. At low soil water content, the water menisci hold the soil particles together very firmly, and a high pressure is required to move them.

Clays and loamy soils respond much more to pressure than sandy soils. The soil density increases almost linearly with the logarithm of pressure applied, as each doubling of the pressure gives a certain step of increase in the soil density. For example, if an increase in pressure from 1 to 2 kg/cm^2 causes the density of a soil to increase from 1.40 to 1.50 g/cm^3, then the application of 4 kg/cm^2 pressure would result in a density of 1.60 g/cm^3, and 8 kg/cm^2 in a density of 1.70 g/cm^3. Each doubling of the pressure has increased the density by 0.1 g/cm^3.

For the same soil with a lower water content, a lower soil density would be achieved at each pressure applied, while the increase for each doubling of pressure would remain similar, although lower; e.g., for 2, 4, and 8 kg/cm^2, the densities may be 1.46, 1.52, and 1.58 g/cm^3.

For a uniform sand the step in density increase for a doubling of pressure may be only 0.05 g/cm^3, or half of that required for the above clay/loam soil, while some poorly structured loamy soils can compact even up to 0.50 g/cm^3 for each doubling of pressure. Such soils are quite unsuitable for horticultural or sportsturf use.

It is important to understand how the above principles apply to the compaction of sporting facilities. If they are used in dry conditions, compaction is low and the soil density is not significantly increased. As the water content of the soil is increased the density of the soil is increased (i.e., it is compacted), with the same usage. As the pressure is increased on the wet soil (the players, horses, or vehicles are heavier) the soil density increases much more. Simply, when facilities are used when wet, they compact more than when they are dry, and the more pressure applied in these conditions, the worse the compaction.

Sand compaction is not affected much by pressure, but it is more sensitive to vibration, with densities being increased by as much as 0.20 to 0.30 g/cm^3 due to applied vibrations.

The effect of an increase in a soil's density on its properties also depends on soil type. While a small increase in the density of a clay soil can be disastrous for its drainage, a similar increase in density of a sand may have only minor effect.

Wheeled traffic on wet soil may compact it to a state where the pore space is filled with water. At this stage no further increase in soil density can be achieved quickly, and a paste of wet soil will flow aside as the wheel proceeds. The result is easily observed in the field, as mounds of soil paste occur alongside the wheel ruts. If the soil is not completely saturated, the wheel rut will appear as a depression only, and the soil surface will remain unaltered alongside the rut.

Particle Movement

The process of particle movement is important because it occurs on all sports surfaces and horticultural facilities when they are used or worked on when wet.

When a soil is saturated, many of the "fines" (i.e., very fine sand, silt, and clay particles) can move in the soil because there are no water menisci to hold them in place. Fast-flowing water may transport even fine sand, while a slower flow will carry only the finer silt and clay particles. Particle movement can happen in a number of situations.

A. When a soil is spread out loosely and subjected to its first heavy rain or irrigation event.

A large number of fine particles that have been detached due to the digging and transporting of the soil will move long distances through the soil. They are carried down the profile by the movement of water through the voids, which are larger in the initial stages due to the loose state of the soil.

This results in a separation of the soil particles, with a layer of coarse sand staying on the surface. The fines may be trapped at some depth in the profile where they may form a seal that blocks further drainage. Otherwise they may move all the way out into drains—if they have been installed—causing dirty drainage water.

In a shrub bed situation, when there is a massive movement of these fines after the first irrigation or heavy rain, these fines may block almost all of the further downward movement of water. When this occurs and the bed ceases to drain, there is very little that can be done to fix the situation other than to start again and replace the soil. Plants will die in this type of situation. Recently the authors observed a major landscaping disaster where more than three thousand plants died in a wet winter and spring from precisely this scenario.

B. When a soil is exposed to excessive heavy traffic in wet conditions.

The soil particles will be squeezed together in a dense packing, with the finer particles almost filling the spaces between the larger ones, as shown in Figure 3.1. Aggregates are disrupted and fill pores between larger sand grains. The arrangement of particles resembles that in a concrete mix, where the aim is to fill all of the voids with fines. In this situation long-range movement of fine particles is not possible, as they will be quickly trapped in the narrow pores.

As a result of this "puddling" process the density of the soil at the top of the profile increases and the hydraulic conductivity decreases. Drainage can almost cease when this process occurs with some silty soils.

When establishing a new soil profile it is important to pack the soil to a moderate density, to avoid the long-range separation of fines. On the other hand, it is important not to compact the soil to such a degree that drainage becomes too slow, aeration is poor, and grass root growth is impeded by soil that is too dense. When the topsoil is initially placed onto the base, it is often fluffed up and very loose. Light compaction, even with light rolling, will ensure that the particles lock into place, and there will be no wholesale movement of fine particles with the first watering or heavy rain. If a depth of topsoil of 8 to 12 in. (200 to 300 mm) is being spread, it should be placed in two layers, with each layer being lightly compacted.

This topsoil should never be spread if it is wet, as this will result in severe compaction. The topsoil, or top layers of soil, should only be rolled or compacted with a roller that applies no more than 150 lb per linear foot (200 kilos per lineal metre). Heavy rolling will only compact the soil to an unacceptable level.

An application of gypsum on the surface of newly laid topsoil will in many cases greatly reduce the amount of particle movement. This can be demonstrated by placing topsoil over a drainage pipe and applying water. There will be a considerable amount of dirty water flow out of the pipe immediately after the first watering, but if this is repeated with a surface application of gypsum the amount of fines washed out will be reduced with some soils. The dirty water is due to a mass migration of fines through the soil.

The process and implications of particle movement and compaction must be understood by those constructing new facilities, whether they be golf courses, playing fields, or other horticultural features, including shrub beds and lawns.

The most significant problems arise when soil is worked when wet. **When heavy machinery is used to shape wet subsoils, immeasurable damage can occur.** The top of the base can be "sealed," and infiltration rates, which may have been good at around 0.04 to 0.08 in. (1 to 2 mm) per hour, can fall as low as 0.04 to 0.08 in. (1 to 2 mm) per day.

This means when the topsoil becomes saturated there is nowhere for the water to go, because it cannot drain downward at a reasonable rate and the facility may remain waterlogged for very long periods of time. This leads to surface damage, prolonged sloppy or muddy conditions, and the death of roots and sometimes of the plants. In shrub beds a sealed base can cause anaerobic conditions, which often lead to the prevalence of root diseases, such as *Phytophthora cinnamomi* and *Pythium* spp., which cause plant death.

In summary, many problems occur when topsoil is worked when wet, resulting in overcompaction, loss of air space, and the slowing of water penetration and drainage. The message is very clear here: **do not carry out construction when the soil or subgrade is wet**. This is not always easy to achieve when contractors are pushing to finish jobs on time, but the consequences of not halting work when the soil is wet will usually be much more costly than continuing. The owner will pay for it for the life of the facility. A facility that will not drain will not grow good grass or plants, and will cost a lot of money each year to renovate. In many cases there is no answer other than to reconstruct.

Bulk Density

Bulk density is a measure that soil scientists use to quantify the density of a soil at different levels of compaction. It is usually expressed as grams per cubic centimetre of soil (Handreck and Black, 1984). As increased pressure is applied to a soil, and there is sufficient moisture present, the smaller particles such as silt are forced into the pore spaces. This process makes the soil more dense and increases its bulk density.

It also decreases the number of pore spaces, thus increasing the number of small pores. Almost always when there is an increase in a soil's bulk density, there is a decrease in its saturated hydraulic conductivity, i.e., its ability to drain.

This concept is important in a sportsground situation because the soil is compacted by playing activities, particularly by players running, jumping, and scrummaging when wet. Compaction causes the soil's bulk density to increase and its ability to drain is reduced, particularly near the top of the profile.

With some loamy soils with a high level of silt, this effect can be quite dramatic. Bulk densities can increase from about 1.5 g/cm^3 to 1.8 g/cm^3 after compaction under wet conditions, and the hydraulic conductivity can drop from maybe 0.25 in. (6 mm)/hr, which is acceptable, to 0.04 in. (1 mm)/hr, which is very poor. This is unacceptable for a sportsturf situation, be it a football field, a racecourse, or a golf fairway. Table 3.1 shows how the bulk density and hydraulic conductivity changed with increased compaction on five common soil types used in sportsground and landscape construction in Canberra.

Table 3.1. Shows the hydraulic conductivity and bulk density at four different levels of compaction for five different soils used in sportsturf and landscaping facilities in Canberra, Australia. The fines are defined as those particles below 0.1 mm.

Soil Type	Use in Sportsturf	Hydraulic Conductivity and Bulk Density	Level of Compaction			
			4 Drops	8 Drops	16 Drops[a]	32 Drops[b]
excellent USGA sand 21% fines only 2% silt	golf greens, bowling greens, sand-based stadiums	HC in./hr	68.0	57.5	50.8	34.8
		mm/hr	1725	1460	1290	884
		BD g/cm^3	1.5	1.54	1.6	1.63
golf sand with too many fines (34%) 10.6% silt	common in golf and bowling greens	HC in./hr	24.8	8.6	3.0	1.0
		mm/hr	629	219	77	25
		BD g/cm^3	1.45	1.58	1.66	1.74

free draining sandy loam 20% fines 9% silt/clay	shrub beds, racecourses, and a few sportsfields	HC in./hr	43.3	23.6	8.5	1.1
		mm/hr	1100	600	217	27
		BD g/cm^3	1.3	1.37	1.48	1.6
silt/clay loam 28% fines 18.% silt/clay	most sportsfields, many racecourses	HC in./hr	29.6	8.6	0.6	0.08
		mm/hr	750	220	15	2.1
		BD g/cm^3	1.33	1.42	1.57	1.75
heavy silt/clay soil 36.8% fines 28% silt/clay	far too many of the above facilities	HC in./hr	41.5	5.0	0.12	0.0008
		mm/hr	1053	128	3	0.02
		BD g/cm^3	1.42	1.58	1.72	1.85

[a] Compaction level approximately equivalent to heavy football use.
[b] Compaction levels approximately equivalent to horse racing.

The hydraulic conductivity and bulk density were measured at four different levels of compaction in the laboratory. These tests were devised by Dr. Bent Jakobsen. (See Chapter 12.)

The soils used in these examples in Table 3.1 are taken from laboratory tests carried out on these soils in Rootzone Laboratories International in Canberra, Australia. It should be remembered that there can be large variations in performance between soils that may appear to have similar particle size distribution. It is therefore very important to carry out compacted hydraulic conductivity and water holding capacity tests on soils to properly assess their performance. Particle size analysis alone is **definitely not** sufficient.

In this test, mild compaction is represented by 4 drops; 8 drops represents compaction affected by, say, teams of small children playing; 16 drops represents approximately the level of compaction applied by sporting use, such as senior gridiron or rugby football; and 32 drops represents approximately the level of compaction exerted by racehorses, i.e., very heavy compaction.

The first sand is an excellent low-compacting sand used for constructing top-quality golf and bowling greens and high-use football stadiums. The second shows how much difference there is in performance if there are too many fines in these sands used for the same purpose. The fines increased by 13 %, but 10.6 % of this was silt.

It can be seen that by selecting the wrong sand to build a green, the hydraulic conductivity has dropped from 50.8 to 3.0 in. (1290 to 77 mm)/hr in the laboratory. The actual hydraulic conductivity one year later in the field may be less than 25 % of these rates.

The extra fines in the second sand tend to fill in the voids, seriously reducing the movement of water through it. These differences are not always apparent to the eye, and there are very few people who can tell these differences by feeling the sands. It is emphasised that they need to be tested in the laboratory before use.

The free-draining sandy loam drains very well, and even under high levels of compaction has still got a hydraulic conductivity of 1.1 in. (27 mm)/hr. While a soil like this drains very quickly, and still drains well at high compaction, it will be very droughty in the summertime. This will make it quite difficult to manage and grow good grass on, particularly cool-season grasses.

For such a soil to be useful in racecourses (its best use), it must have a water holding capacity at one metre suction of at least 12 % by weight. Such soils are difficult to find unless they possess a naturally high organic level, or the clay content is high, compared to the silt content.

A silt/clay loam can still perform well, as shown by the example, which had a hydraulic conductivity of 0.6 in. (15 mm)/hr at 16 drops. This soil would be suitable for building good sportsfields and golf fairways. The tests show that there will still be adequate drainage even under high usage. This type of soil will have a water holding capacity of around 15 %, which is sufficient to grow good grass, shrubs, and flowers.

This soil is ideally suited for facilities that have a good surface slope to shed water, and for the production of a good dense grass sward. However, it will compact when wet.

The heavy silt clay soil is a poor one for use in a sportsfield, as it will compact badly, and the hydraulic conductivity will drop quickly as the fines are redistributed. A hydraulic conductivity of below 0.2 in. (5 mm)/hr at 16 drops is unacceptable for sportsground use.

Unfortunately, there are a large number of facilities, including golf fairways, that are built from this type of heavy material. There is little that can be done to improve the drainage of these soils once this initial choice has been made.

When the top of the soil profile becomes saturated, these finer particles tend to "float," and are easily pushed into voids when pressure is applied. Enormous pressure can be applied to wet soil by football use—such as rugby scrums—and by horses. This pressure moves the fines, forcing them into the voids, which causes compaction. Grass roots have an average diameter ranging from 60 μm to 250 μm (Adams and Gibbs, 1994), and as the compaction process reduces the number of larger pores, there are fewer pores above 60 μm in diameter remaining in the soil. This reduction in air space—and the resultant reduction in hydraulic conductivity—makes this compacted part of the soil profile less favourable for root growth and so tends to restrict it. Hence, there is often shallow-rooted grass in compacted areas.

If shrub beds are compacted, the growth of plants is usually poor, and waterlogging can occur in periods of prolonged wet weather.

With the loamier materials, there can be very large movement of particles in some cases, particularly if there is a high percentage of silt in the fines. Silt particles tend to be the easiest moved, as often clay particles bond together and act as much larger particles. Silt and very fine sand particles are almost always free to move when the soil becomes saturated.

The most vulnerable of all are manufactured soils, which are mixtures of very silty river soil with sands. These are regularly sold as sandy loams. The silt moves very quickly at the first watering, and can clog up the whole profile. Once this occurs there is little that can be done to remedy the situation. Facilities made from these soils will always drain poorly and be unsuitable for good grass or other plant growth.

The message is loud and clear—**beware of silty sands**. They often look good, feel good, and are promoted by soil yards as good sandy material.

A simple test to see how much silty material there is in a topsoil is to place some soil in a small nursery pot, e.g., a 3-in. (75-mm) tube, and drop it four times from a height of about 4 in. (100 mm) onto a firm surface. Then add water to the top of the pot. If there is a large amount of dark dirty material washed out of the bottom of the container the soil probably has a high silt content. The more stable the soil, the less dirty water will be flushed out the bottom of the tube.

Note

For the purposes of these discussions, *clay* consists of particles less than 0.002 mm; *silt* particles are those between 0.05 and 0.002 mm; and *very fine sand* is between 0.1 and 0.05 mm. The combination of these three fractions is referred to as the "fines."

CHAPTER **4**

The Capillary and Lateral Movement of Water in Soils

Chapter 2 considered how water enters the soil and moves downward through it, and also discussed gravity, which is the main force affecting this movement of water. There is, however, another process whereby water is moved in soils, and this is by capillary action, or capillary forces.

Before we embark later in this chapter on the broader issue of lateral movement of water in soils, this very important area of capillary action needs to be discussed in some detail.

The Capillary Movement of Water in Soils

As explained in Chapter 2, water molecules are very strongly attracted to the surfaces of the pores, i.e., the soil surfaces. As the pores become narrower this attraction of the water molecules to the surfaces of the soil particles becomes increasingly important.

This attraction can be even stronger than the force of gravity pulling the water molecules downward. In fine pores the adhesive forces attracting the water molecules at the edges of the menisci will in fact "pull" the menisci upward against the force of gravity, which is attempting to pull them downward. It can also pull these menisci sideways, and in fact in all directions away from a water source in the soil. (See Figure 4.1.)

This movement of water in the fine pores by these forces is called *capillary action*.

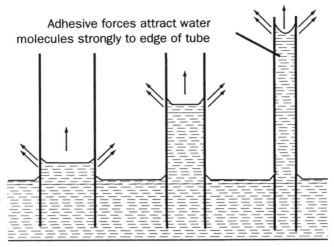

Adhesive forces attract water
molecules strongly to edge of tube

Water molecules at the edges of the menisci are attracted
so strongly by the sides of the tube it actually
moves them up the tube against gravity.

Figure 4.1. Menisci rising in the tubes against gravity. The smaller the tube diameter, the further the menisci are "pulled" up by the capillary forces, i.e., the adhesive forces on the tube surfaces attracting the water molecules at the edges of the meniscus.

The narrower the pores the further the water will move up, along, down, or away from the source of the water. Most people have heard of the capillary rise of water in soils, but its movement in the other directions is not widely written about. If the capillary sizes are the same, then water will move further downward away from the source of water than it will move upward because the downward movement is being assisted by gravity, whereas the upward movement has to occur against gravity.

Capillary movement of water in soil occurs after rain or irrigation, or from a source of water within the soil, such as a leaking pipe or a spring. It happens particularly when there are some areas of soil which become saturated with areas adjacent to them which are dry. This dry soil may be under, beside, or even above this saturated or very wet zone, depending on where the source of water originates. (See Figure 4.2.)

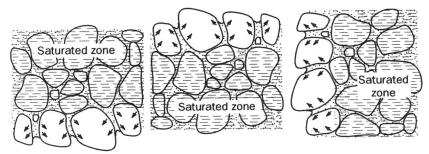

Water being drawn downward, upward, and sideways
away from saturated zone by forces of adhesion.

Figure 4.2. Menisci being pulled up, across, and down through narrow pores by the adhesive forces on the surfaces of these pores attracting water molecules at the edges of the menisci.

Remember that the water only moves as a result of capillary forces until the soil is near field capacity, and that it ***never*** moves through the large pores by this means.

The movement of water laterally or upward by this means is therefore very slow. When it stops at field capacity, it has rarely moved more than 4 in. (100 mm) from its source. Thus, we can conclude that *the capillary movement of water in soils is very slow, and this movement is over very small distances.*

The relative distances that water moves by capillary action depends largely on the type of soil, in particular its texture, or more precisely the number of small pores. The capillary movement of water in a coarse sand will be almost negligible; however, water will travel quite perceptible distances through a silty clay soil.

Hence, we may conclude that the contribution of capillary movement of water has virtually no significance in the removal of water in the context of horticultural or sportsturf drainage.

The Lateral Movement of Water in Soils

A basic understanding of the contribution of both the principles of the downward movement of water and of capillary processes is essential in the understanding of the quite complex process of how water moves laterally (sideways) in soil.

The lateral movement of water in soils is probably the area of drainage where people have been most influenced by misconceptions, myths, half truths, and wrong assumptions. We even find that it is wrongly taught in horticultural institutions, and many textbooks are inaccurate on this subject.

This subject has traditionally been oversimplified in an attempt to teach it to people who do not have a background in physics and mathematics. In doing so, the core principles of physics have been overlooked, and often very inappropriate assumptions and analogies have been made.

Most of these inaccuracies fall into the area of the lateral movement of water in soils. This is because the lateral movement of water in soils is quite complex, and without a firm grasp of soil physics it is difficult to make assumptions on how water will behave in soil. Wrong assumptions have been made, and these have been written about and taught—probably for most of this century. These mistakes are not confined to Australia, as this is a worldwide problem in horticulture and sportsturf. Most of these assumptions have been made about the speed at which water moves laterally in the soil and the amount of water that will move laterally in a given time.

Once the wrong assumptions have been made on these two parameters, the mistakes follow, particularly in the assumptions on what subsoil drains will do, and at what spacings they should be installed to achieve a particular end. Consequently, we see subsoil drains spaced 10 m apart on playing fields that have been constructed using ordinary soils, which compact (this is about 99 % of all soils). We also see subsoil drains at the edge of 6- to 8-foot (2- to 3-m)-wide flower beds. These scenarios will never work, because it may take weeks for water to move 6 to 8 feet laterally through the soil to reach these drains.

Newly landscaped areas behave quite differently from agricultural land, where the profile may not have been greatly disturbed for a very long period, even up to hundreds or thousands of years. The requirements of drainage systems are different for both of these situations. Unfortunately, almost all the literature and research has been done for agriculture, and the transfer of this information across to the contrived situation of urban horticulture and sportsfields does not always apply.

With this background in mind, the authors have endeavoured in this book to give the reader a better understanding of how water moves laterally in soil by presenting it in the "soil physics for beginners" form in plain English.

Hopefully, as a result, the reader will be able to understand how subsoil drains can be used to remove soil water in typical horticultural and sportsturf situations. They will also be able to do the basic calculations themselves in order to achieve the drainage rates required. In addition, readers should have some idea of what to expect from tees, fairways, sportsgrounds, flower beds, etc., that do not have subsoil drains.

Rising Water Table

Wet areas can be caused by a rising water table due to below-ground water flowing into the area. This problem is most easily recognised early in a dry season, when the surrounding areas are fairly dry but the affected area is still wet. The wet area is more likely to be located in the middle of a slope rather than at the bottom of it. In the latter case, i.e., at the bottom of the slope, the problem is more likely to be a surface water problem.

Water belowground flows along seams of gravel in the soil profile. Such a seam of gravel may be covered with an impermeable clay soil of variable depth, and where the seam is close to the surface the water may break through and create a wet area. Water will enter the seam far up on a hillside, often several hundred yards or even miles away, and the pressure in the seam can be considerable. However, to deliver sufficient amounts of water to cause any problem, the gravel in the seam must be coarse and clean.

The water source can be located with the use of a *piezometer,* which is a thin steel tube with a pointed tip and holes in the wall behind the tip for entry of water. The piezometer is driven into the ground, and when a water source is met, water will rise up in the tube.

The best remedy to the problem is—as for surface water—to prevent the water reaching the playing surface or other facility. The pressurised water can be intercepted in the gravel layer outside the facility further uphill. Dig a trench through the layer and into the soil below and install a drainpipe, in accordance with the procedures discussed in Chapter 9. This action will take the pressure of the stream and the wet area will disappear without the need for drainpipes to be installed in the wet area itself.

Natural Deep Drainage

When surface water runoff has ceased, a small amount of free water is left in local depressions on the soil surface and in the larger noncapillary pore spaces in the soil. Such water must be drained down through the profile to ensure proper aeration of the topsoil and the root zone of the grass or plants. This may happen by natural drainage downward through the subsoil, or into drainpipes if these are installed.

Water movement through topsoil is much slower than flow over the surface, because in soil it must move through a narrow, tortuous pore system. Flow through the subsoil is very much slower because, as a rule, the average pore sizes are very much smaller than those in the topsoil. The rate at which water drains down through the subsoil is of great importance, because once the topsoil becomes saturated the rate at which any further drainage occurs from that topsoil will be the same as the drainage rate of the subsoil.

Before considering the installation of drains, the properties of the subsoil must be evaluated. For certain subsoils the natural drainage rate may be as high as 0.04 in. (1 mm) or more per hour, and in these cases drainage problems are greatly reduced.

In a study of 20 sites near Canberra, ten of the subsoils had a hydraulic conductivity of more than 0.04 in. (1 mm) per hour, which could provide adequate drainage for most sportsfields and shrub beds. Only one soil had a hydraulic conductivity of less than 0.04 in. (1 mm) per day. In this case, because the natural rate of drainage was so poor, any facility constructed on that subsoil would have had severe drainage problems.

In many playing fields, racecourses, and golf course fairways, water movement down through the subsoil is very restricted because the soil was compacted during construction. Another cause of slow drainage is soil with a high silt content and poor structure. These soils often occur in low floodplain areas, and sporting facilities are frequently constructed on these sites because they are unsuitable for housing and other development.

For sites that have good natural drainage, efforts should be made during construction to maintain this condition and eliminate the need for drainpipes. *All grading and levelling of the base should be done while the soil is so dry that it crumbles and the surface does not become smeared.*

An easy way to check if the soil is too wet for working is to take a lump of soil in the hand and try to mold it. If the soil can be molded at all, then it is too wet for work involving high pressure, such as grading and subsoiling. If the soil is so wet that it can be rolled out into a string that is less than 0.12 in. (3 mm) in diameter without crumbling, the soil is too wet even for light working.

For areas that receive a deep layer of fill, it may not be possible to stabilise the base sufficiently to prevent future sinkage if working with dry soil. One solution to this problem is to fill low areas well in advance of subsequent work and allow the soil to settle naturally before final grading of the base.

After the base has been graded and before placing the topsoil, it should be ripped to a depth of at least 12 in. (300 mm), and preferably deeper, so as to eliminate any sharp interface between the base and the topsoil. Application of gypsum at the rate of 100 lb/1,000 ft^2 (500 g/m^2) after ripping will stabilise the structure and increase the hydraulic conductivity of clay soils, at least temporarily. This treatment is often desirable because the soil structure has been damaged by the construction or landscaping work and drainage rates have been reduced. Years later, when all the gypsum has been leached out of the soil, its structure will have stabilised again and the effect of the gypsum is no longer needed.

How Water Moves Laterally in Soils

As mentioned earlier, one very important fact to remember is that water moves laterally, extremely slowly through the soil. This rate may be from ten times, to more commonly in excess of one hundred times slower than it moves vertically down through the soil.

There are four aspects relevant to sportsturf and horticultural drainage that are fundamental in the understanding of the lateral movement of water in topsoil, and the rate at which it moves in that soil. These are:

1. The soil—or a zone of the soil—must be saturated, and there must be "free water" (see explanation in Figure 5.2) in this saturated zone before *any* water will move laterally, other than tiny amounts moved by capillary action.
2. If this saturated free water zone is sitting on a slower-draining base, for water to move significant distances later-

ally from this zone it must be removed from the profile by some means, such as subsoil drains, the edge of a shrub bed, a dish drain, etc.

3. The slope on the base will play some part in the lateral movement of water. This slope on the base will determine the component of gravitational force (usually a small fraction) that contributes to the lateral movement of this water.

4. When water moves sideways (rather than downward) through the soil for a considerable distance, it must be removed at the end of the slope for this movement to continue; otherwise it will pond. The rate at which it moves is constrained by the cross-sectional area of the saturated free water zone of the soil through which it is moving. This is usually much less than the area on the surface through which this volume of water entered the soil.

Lateral Movement of Water When a Saturated Zone Does *Not* Reach the Slower-Draining Base

Where there is a section of the topsoil that is saturated but does not extend right down to the base (Figure 4.3), the rate at which the water moves **down** through the soil is similar to the saturated hydraulic conductivity of the topsoil. Capillary action in the soil in a very small band below the saturated zone will slightly increase the rate of the downward movement of water and is called the wetting front (Yong and Warkentin, 1975).

In the situation where the whole surface of the soil is wetted by rain, there will be virtually no sideways movement of water, because there will be a section of the profile that will be saturated to approximately the same depth right across the whole area.

The major difference in the amount of variation of this saturated zone will be the variation in the texture or pore space distribution between adjacent areas of topsoil. For example, compacted areas will not become saturated to the same depth as adjacent less-compacted soil, simply because they will have a lower hydraulic conductivity because they have fewer large pore spaces.

This is quite significant, as it means that compacted areas of a sportsground will not be wetted to the same depth as adjacent less-

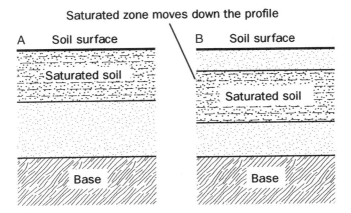

Figure 4.3. Saturated zone of the soil in relation to the slower-draining base. In A the saturated zone is near the surface, and in B it has moved down toward the impermeable base.

compacted areas by a rainfall or irrigation event. This often leads to shallower grass roots developing on these areas, which are consequently more easily damaged by heavy wear. They also have higher capillary fringes. (See Chapter 5.)

Under what circumstances does water move laterally in the situation where the saturated zone has not yet reached the base?

Let us consider the situation where there is some object, even as small as one square yard of plastic, on the surface of an ordinary football field or golf fairway. When rain falls onto the grassed area, a zone below the surface becomes saturated and water begins to move downward. No water has entered the soil covered by the plastic, hence the soil will be dry directly below the plastic and adjacent to the saturated zone below the grass.

In the types of soils that are generally used for sporting surfaces, where the depth rarely exceeds 10 in. (250 mm), the sideways movement of water under the above conditions would rarely exceed 4 in. (100 mm). This means that the angle of the capillary wetted soil in Figure 4.4 would never exceed 45°, and would more likely be 75° to 80°.

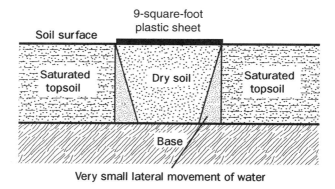

Figure 4.4. A shallow soil profile of 10 in. (250 mm) of topsoil with an area covered by a 9 ft² (0.84 m²) of plastic. Water moves downward to the base and there is very little movement of water sideways before the saturated zone reaches the base.

How and why does any water move sideways in this scenario?

In the saturated zone all the large pores are full. As this zone moves down through the soil, some of these large pores will be directly connected to large pores in the adjacent dry soil that are not orientated downward, and water flows into them. Once they become full, gravity will begin to pull the water directly down from them through the smaller pores connected to them.

From all these large pores on the interface between saturated and dry soil, there will be capillary movement of water laterally. This will have the combined effect of moving the water predominantly downward, but deeper down in the profile an increasing amount of water will "leak" sideways. Remember, however, that this sideways movement is really very small.

In a profile 10 in. (250 mm) deep or less, the very small distances that water moves sideways before it reaches the base are of very little significance in the overall drainage of water in sportsturf. This sort of situation is of great importance when we are considering what happens under the ground when the surface is watered by an irrigation system.

With most irrigation systems there is not uniform water distribution across the whole area. Even the very best designed systems have differences of water distribution that vary by up to a factor of two. It is not uncommon for small areas to receive as little as 0.24 in. (6 mm)/hr, and other areas to receive 0.55 in. (14 mm)/hr, when the system delivers an average of 0.39 in. (10 mm)/hr. In many systems of poorer design, and/or poorer sprinkler distribution, these differences may vary by as much as a factor of 10, e.g., 0.12 to 1.18 in. (3 to 30 mm)/hr.

What happens under the soil below these differing surface water applications?

In actual fact, this is much the same as the previous example of the one square yard of plastic sheet laid on the surface of the sportsfield, but instead of no water falling on an area, there will be less falling on the adjacent area. This results in the depth of the saturated area—and ultimately the zone of the profile that becomes wetted—remaining very much in proportion to the amount of water that fell on the surface of the soil.

If the application of water to the surface is uneven, then the depth to which the water penetrates the profile will be equally uneven. This is provided that the saturated zone does not reach an impermeable or slower-draining base. This concept is extremely important when understanding the impact of uneven irrigation distribution on root depth and grass growth.

Uneven irrigation on the surface does not even out under the soil by the lateral movement of water within the soil—unless one area is flooded for a prolonged time; however, this is a completely unacceptable irrigation practice for sportsturf and general horticulture. (See Figure 4.5.)

This situation is well demonstrated by the fact that, even after heavy watering in the summer, those areas receiving less water do not wet to any great depth, whereas the adjacent soil receiving the heavier watering will wet to a considerable depth. These differences are expressed by differential grass growth, and often even by different species being favoured by these differing wetting patterns, e.g., annual bluegrass in the very wet areas and Bermudagrass in the drier areas.

If water moved sideways in the soil at the rates that some people would have us believe, most of this uneven distribution of water would

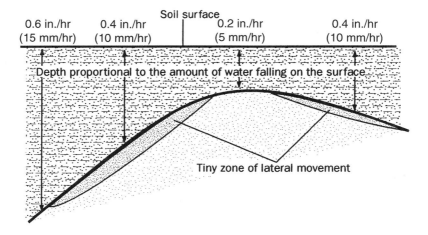

Figure 4.5. Different depths of saturation of the topsoil after or during an irrigation event, due to uneven water distribution on the surface.

disappear as water magically flowed laterally, levelling out the saturated wetting front. Sadly, this is not the case.

Remember that water moves laterally in soils *extremely slowly*, and in most situations it can be ignored for the purposes of drainage.

Lateral Movement of Water When a Saturated Zone Reaches the Slower-Draining Base

The above sections have dealt with the downward movement of water in soil driven by gravity, and the lateral movement of water by capillary action before a saturated zone reaches a slower-draining base. Once the saturated zone reaches a slower-draining base a whole new set of conditions apply, and the movement of water laterally becomes much quicker, although still relatively slow.

This situation is very common in sportsturf and horticulture, as almost all sportsgrounds, golf course fairways, racecourses, parks, shrub beds, etc., consist of topsoil of varying depths being laid over a subsoil base, which often has a very slow drainage rate, and almost universally is slower than the drainage rate of this overlying topsoil.

Usually the depth of the topsoil is shallow, ranging from 3 in. (75 mm), to, in rare cases, 12 in. (300 mm). The depth of this topsoil layer is crucial in the whole drainage picture. In particular, it is highly relevant to how quickly the soil profile becomes saturated, and how quickly a saturated zone reaches the very slow-draining base. It is also crucial in relation to the depth of the capillary fringe. (See Chapter 5.)

The rate at which the base drains is also crucial, as even a drainage rate of 0.04 in. (1 mm)/hr can be significant in removing water from the surface of the soil, and in many cases it will be a lot faster than the subsoil drains that have been installed. Traffic from earthmoving machines used for the shaping of the base often causes so much compaction to the subsoil that it becomes almost impermeable, which can adversely influence the complete drainage of that area.

Before the "saturated" zone reaches the base no water can move sideways, as described above. Once it reaches the base there has to be a buildup of *"free water,"* i.e., water at nil suction, before any water can move sideways. This saturated zone of free water is the only water that can be moved sideways by gravity, as the saturated zone above it is called the **capillary fringe.**

Water is moved down through the soil profile by gravity, but only the free water is moved sideways in the soil by gravity. This free water only moves sideways if there is a component of gravity pulling it, which is usually achieved by the slope on the base* causing a **head** of water, so that the water is being pulled down the slope by gravity. The steeper the slope, the greater the effect of gravity.

For example, if the slope on the base is 1:100, the water is only being pulled sideways across that slope at one-hundredth of the rate that it can move downward. If the slope is 1:5, it will move sideways one-fifth as fast as it can move downward, or one-fifth of the topsoil's hydraulic conductivity.

If the hydraulic conductivity of the soil were 0.79 in. (20 mm)/hr, the maximum rate that water could move sideways—even on this very steep slope—would be 0.16 in. (4 mm)/hr.

If the slope on the base is only 1:100, the maximum rate at which water can move sideways is $0.79 \times 1/100 = 0.008$ in. (0.2 m)/hr.

---

* Water will move laterally toward a drain or some other situation where it is being removed, even though the base may be flat. Under these circumstances the gradient on the depth of the saturated zone, as it is being depleted, provides the gravitational component. (See Figure 5.7.)

So, the reader can see that even on a steep slope water cannot move very quickly sideways. On the more common gentle slopes, water moves sideways very slowly indeed. ***Water must be removed at the end of the slope for this process to proceed.***

When the saturated zone reaches the base and there is a buildup of free water, this water begins to move laterally down the slope. For this process to continue, water must be removed from the system by some means. This may occur with drains, the edge of a flower bed may "weep," or weep holes may have been placed at the base of a wall. If this does not occur, an equilibrium will be reached. Often this results in water ponding on the surface and swampy conditions at the end of a slope, or against an impermeable obstruction such as a wall. (See Figure 4.6.)

There can be a buildup of hydrostatic pressure under these conditions, and this pressure can cause the collapse of walls and land slips. There will be free surface water at the end of the slope if there is no drainage to remove it.

The rate at which water "flows" laterally is influenced and restricted by three factors:

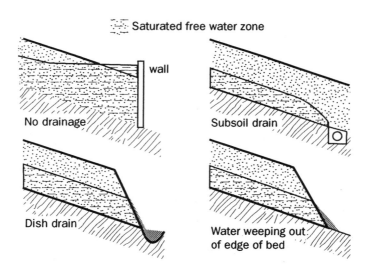

Figure 4.6. Water needs to be removed at the end of the slope, otherwise it will pond at the lower end.

a. depth of the saturated free water zone in the topsoil
b. hydraulic conductivity of the topsoil
c. slope on the base

How does the depth of the saturated free water zone in the topsoil influence the rate of lateral movement of water?

Once the saturated zone has reached the base and free water has built up, the depth of this saturated free water zone is the most important factor influencing the rate at which water moves laterally in soils.

The ratio of the cross-sectional area of the soil through which the free water travels laterally to the area of the soil surface over which the rain was collected is very important in understanding the whole picture of how water moves laterally in soils over a slow-draining base.

When rain falls on the surface of the soil for long enough, it enters it and will saturate a zone of topsoil. When the rain ceases, the saturated zone will continue to be drawn down through the profile by gravity at the rate of the saturated hydraulic conductivity of the topsoil.

Once it reaches the base, and free water builds up directly above the base, it will begin to move sideways toward a drain, or other situation, that is constantly removing water. If the cross-sectional area of the saturated free water zone is smaller than the area over which the water was collected—which it almost invariably is—then the rate at which the water is able to move sideways is restricted by the ratio of:

$$\frac{cross\text{-}sectional\ area\ through\ which\ the\ water\ is\ moving}{area\ over\ which\ the\ water\ was\ collected}$$

Example 4.1. Let us consider a topsoil 12 in. (300 mm) deep, with a hydraulic conductivity of 1 in. (25.4 mm)/hr. The saturated zone has reached the base, and it has built up to 9 in. (230 mm) of free water. If we consider that the surface area over which the water was collected is 9 ft^2 (0.84 m^2), and this water is moving laterally through the soil toward a drain that will remove it, this water will only be moving through a pipe with a cross-sectional area of one-quarter the size of the area over which it was collected, i.e., 3 ft × 0.75 ft = 2.25 ft^2 (0.21 m^2), 2.25 ft^2 /9 ft^2 = 0.25, or one-quarter the area. (0.84/0.21 = 0 25).

The maximum rate that water could move laterally was 1 in./hr ×
0.25 = 0.25 in. (6.4 mm)/hr. This rate will be *further reduced by
the gradient on the base,* which is explained later.

Example 4.2. Using the same profile as in Example 4.1, and assuming that the pore space through which the water can travel (i.e., the
air-filled porosity, which is the pore space available between field capacity and saturation), is 15%, the following estimations are made.
These estimates are simplistic and assume that the soil profile is near
to saturation at the time of the rainfall event, but can give some indication of what happens in the field.

It would have taken in excess of 1.35 in. (34 mm) of rain to form
the 9 in. (229 mm) of free water zone, i.e., 15% of 9 in. = 1.35 in.
(15% of 229 mm = 34.4 mm).

We can assume that the rain that entered moved down through
the soil at the rate of 1 in. (25.4 mm)/hr, i.e., the hydraulic conductivity of the soil. It would have taken about one and a half hours to
saturate the top 9 in. (229 mm), and a further three-quarters of an
hour for it to reach the base.

From Figure 4.7, the water that travelled down through a "pipe"
that was 9 ft^2 (0.84 m^2) in cross-sectional area reached the base built
up to a depth of 9 in. of free water. If this water is to move sideways,
it must now move through a "pipe" that is only one-quarter of that
area, or 2.25 ft^2. The rate of sideways movement would have been
reduced from 1 in./hr to 0.25 in. (25.4 to 6.4 mm)/hr by the restriction of the "pipe."

The depth of the topsoil layer itself is a limiting factor in this process. If the depth of the topsoil is only 4.5 in. (115 mm) and the soil has
free water to the surface, then in the above example water could only
move sideways at a rate one-eighth the rate it can move downward.

If the topsoil depth is 1 ft (300 mm), there is the potential for
water to move toward drains much more quickly, because the saturated free water zone can be deeper.

The depth of topsoil, and more importantly, the depth of the saturated free water zone of this topsoil, is crucial in determining how
quickly water will travel laterally in soil toward drains. The depth of
the free water zone in the topsoil determines how much of a reduction in the rate (from the soil's hydraulic conductivity) of water moving sideways occurs.

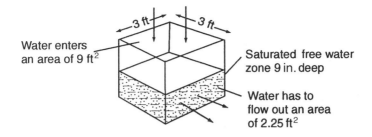

Figure 4.7. Water collected from a surface area of 9 ft² to move laterally through a "pipe" with a cross-sectional area of 2.25 ft², which will constrict the flow rate to one-quarter of the hydraulic conductivity of the topsoil.

Quite obviously, from this simple example it can be seen that the shallower the topsoil, the closer drains have to be spaced.

How does the saturated hydraulic conductivity of the topsoil affect the movement of water laterally in the soil?

These aspects have been discussed earlier. The rate at which water moves sideways through saturated soil is directly proportional to the rate at which it moves downward through saturated soil. In other words, the smaller the pore spaces in a soil, the slower water will move sideways. Water will move much faster through a coarse sandy material than it would in a silty clay soil.

How does the slope of the base influence the rate of lateral movement?

When the saturated zone reaches the base, and free water begins to flow toward a drain, gravity is the force responsible for this sideways movement of water. It is possibly better explained by the weight of the water in the saturated free water zone pushing the water down and along the slope. Of course the greater the depth of the free water zone, the greater the head of water. The greater the slope on the base, the more this adds to the increase in the movement of water, because it increases the height of the head.

This rate at which water moves sideways in soils will be further decreased from the rates calculated in the above example when the

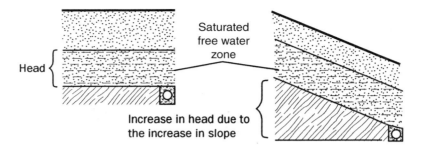

Figure 4.8. The increase in the slope adds to the height of the head, which increases the pressure forcing the water downward.

slope of the base is taken into account. For example, using the same data as for the above examples, if the slope on the base were 1:100, then the rate of sideways movement in this example would not be 0.25 in./hr but 0.25 in./100 = 0.0025 in./hr (0.064 mm/hr).

In simple terms, in this soil the water is reaching the base (with a slope of 1:100) at a rate of 1 in./hr and moving sideways at a maximum rate 400 times slower at 0.0025 in./hr.

Even if the slope were very steep at 1:5, the rate would only be 0.05 in. (1.27 mm)/hr. This is extremely slow, particularly as this soil is a sandy loam with a good drainage rate of 1 in./hr, and is 1 ft deep with a free water zone of 9 in. Even under these very good conditions with a steep slope and a good-draining topsoil, water would take many days to move 1 ft sideways. (See Figure 4.8.)

On normal shallow soils and gentle slopes, which exist on most golf courses, football fields, and racecourses, the rate of lateral movement will be much less than discussed in this example, simply because the depth of topsoil is rarely 1 ft (300 mm).

THE LATERAL MOVEMENT OF WATER IN SOILS IS EXTREMELY SLOW.

CHAPTER **5**

The Capillary Fringe and How Water Gets into Drains

One of the most important aspects of drainage is the mechanism by which water moves toward and enters subsoil drains. As outlined in Chapter 4, the soil has to be saturated and "free water" present before water will move sideways in it, and when it does, it moves very slowly. All water has to move sideways in the soil before it reaches subsoil drains, except that water directly above the drain itself. The very low rates of sideways movement previously discussed are the rates that have to be dealt with when considering how water gets into drains.

There is, however, another phenomenon operating within most horticultural soil profiles. This is the *capillary fringe,* which was briefly mentioned in Chapter 2. The capillary fringe makes the effectiveness of subsoil drains very small indeed, if their purpose is to drain the water out of the top of the soil profile.

Capillary Fringe

When a soil drains in its natural state there is often a deep topsoil layer, and the top of the profile drains down at an even rate. Usually the top 4 to 6 in. (100 to 150 mm) of the profile will reach field capacity quickly after rain ceases, because there was "room" for the water from the top of the profile to be steadily accepted into the empty pore space in the profile below.

With most sportsfields and manufactured horticultural features the situation is different. There is usually a relatively thin layer of topsoil—4 to 12 in. (100 to 300 mm)—placed over a heavier clay base. This creates a situation where the topsoil, or part of it, will become saturated by heavy rain. The result is a zone of saturated soil at the interface between the topsoil and the base, because the water is reaching the base faster than the base can absorb it.

If the base is dry it will apply suction to the topsoil above, and this will act to remove small amounts of water from the topsoil. Provided the amount of water reaching the base is small, this water will be removed steadily at the rate of the hydraulic conductivity of the base, and the top of the profile will drain to field capacity and remain mechanically stable.

When there is a significant amount of water reaching the base quickly, as is the case during heavy rain, a different situation arises. A zone of saturated topsoil develops at the interface of the two layers, and this will only be depleted at the rate of the hydraulic conductivity of the base. This zone will remain saturated for a long period of time, and the height of this saturated zone can sometimes extend right to the surface, particularly with shallow profiles.

As this saturated zone deepens above the base, a situation develops where there are (for the sake of simplicity) two types of saturated soil. In the saturated topsoil immediately above the base a layer of free water develops. This water is no longer under the influence of suction, and can move sideways in the soil. The zone of saturated soil (some physicists call it quasi-saturated, or the *transmission zone*) above the free water zone is called the *capillary fringe.* The water in the capillary fringe *cannot move sideways in the soil,* other than in small amounts (which for the sake of our understanding can be ignored). Water in the capillary fringe of a USGA sand will move much greater distances than in ordinary soils. This may have to be taken into account when building perched water table profiles on steep slopes.

At the interface between the free water zone and the capillary fringe there is a state described by the physicists as *nil suction.*

(The water in the capillary fringe is a perched water table, and is dealt with in detail in Chapter 6.)

To illustrate this, take a 1 in. (25 mm)-diameter tube that has the bottom covered with an open gauze to prevent the soil from falling out and to allow for free drainage, and fill it with soil. Lightly compact the soil by dropping it a couple of times from about 6 in. (150 mm). Fill the tube with water from the bottom by immersing it in a container with a depth of water greater than the depth of the soil in the tube. Remove and place on a sink to drain. If, after all the water is allowed to drain from the tube, the moisture content of the soil is taken at various levels down the tube, the following graph can be made, plotting moisture content against depth. This is essentially what happens when a moisture release curve—or water retention curve— is made. (See Figure 5.1.)

In the bottom section of the tube there is a zone where the soil is still saturated, but no water can move sideways from it, because the force of gravity needed to pull it sideways is not as strong as the meniscal forces at the top of the large pores that are holding this water. Sometimes this water is referred to as "suspended water," but we call it the *capillary fringe.*

The capillary fringe is of great significance, because the water in this zone is not able to move sideways, and therefore it cannot move into drains.

When all of the water that can be drained from the tube has done so, an equilibrium has been reached. There is considered to be nil suction at the very bottom of the tube.

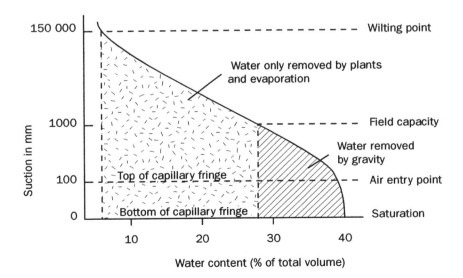

Figure 5.1. A moisture release curve for a sandy loam soil.

If at this stage the bottom of the tube is sealed and more water is added from above, the extra water will cause a zone of free water to develop at the bottom of the tube. This water does have the ability to move sideways. As more water is added to the top of the tube the depth of this free water zone increases. The point where there is nil suction also rises, as this is the interface between the capillary fringe and the free water zone. (See Figure 5.2.)

This is approximately what occurs in the field situation, where, as the zone immediately above the base becomes saturated, more rain will cause the formation of a free water zone directly above the base. If there is a gradient on the base, or if there is a drain that is taking water away—which in effect creates a gradient—then water can move sideways from this free water zone.

There is a point on the moisture release curve called the air entry point. This is the point above which the pull of gravity is strong enough to break the surface tension on the top of the menisci of the large pores, and the water in these large pores drains down the profile.

The further up toward the surface one moves, the stronger the pull of gravity. Consequently, more water is pulled down out of increasingly small pores. This continues until the soil reaches field capacity, where no more water can be pulled down through the profile because the forces of adhesion and surface tension are equal to the

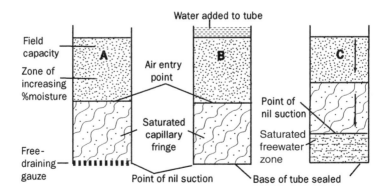

Figure 5.2. In tube A, drainage has ceased through a permeable base. In tube B, the base is sealed, and extra water is added to the top of the tube. In tube C, there is a rise in the point of nil suction, and there is a zone of free water developing at the bottom of the tube.

force of gravity. This is why the water content decreases from the air entry point to field capacity, as shown by the sloping line on the curve in Figure 5.1.

In the area of the curve between field capacity and wilting point the water can no longer be removed by gravity and has to be removed by plant transpiration and by evaporation.

Provided the soil profile is deep enough, the surface can drain back to field capacity (and mechanical stability) while there is still free water at the top of the base. The air entry point must be below the soil surface, otherwise the soil profile can remain completely saturated for long periods.

As the free water is moved down into the base, the top of the capillary fringe moves down at the same rate.

If a hole dug into the soil reaches the saturated free water zone, water moves sideways and enters the hole because there has been sufficient gradient created by the depth of the hole for water to move sideways into it. (See Figure 5.3.)

Even though the soil immediately above the saturated free water zone is also saturated, *no water* will move sideways into the hole. In addition, it must be emphasised that no water can move sideways from

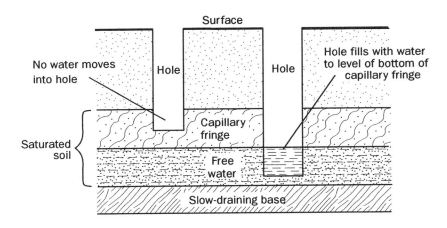

Figure 5.3. Water moves from the saturated free water zone into a hole dug down into it, but no water moves into the hole when it is dug into the capillary fringe.

the capillary fringe into subsoil drains. This is because the water is held in the pores by the combined forces of the surface tension of menisci and the adhesive forces of soil particles, which together are stronger than the component of gravity that is trying to pull water sideways.

The height of the capillary fringe increases as the average pore size of the soil decreases. This means the height of the capillary fringe in loam soils may be as much as 8 in. (200 mm). In most sportsturf situations this will mean that the topsoil will remain saturated to the surface as long as there is any free water at the top of the base.

The other serious implication of the capillary fringe is that areas of compacted soil have less pore space. Hence, in these areas the capillary fringe will be higher. This simply means that these soils will stay wetter longer than the adjacent less-compacted topsoil. This is another reason why compacted areas continue to perform badly compared to adjacent uncompacted areas.

In situations where the soil profile is deep enough, there will be a saturated free water zone above the base, a capillary fringe, and the soil above the top of the capillary fringe, which can be at, or nearing, field capacity.

It is important to understand that the only water that can move into the subsoil drains is from the free water zone. No water from the capillary fringe can move sideways toward the pipe even though this zone is saturated. In fact, what is occurring is that the drains are lowering the depth of the free water and the capillary fringe is moving down at the same rate. (See Figure 5.4.)

As the water moves into the drain, the height of the free water zone diminishes, and so too does the gradient from the drain to the top of the midpoint between the drains. As this gradient diminishes, so too does the rate of movement of water into the drains. This is because the component of gravity that is pulling the water sideways is diminished as the height of the free water zone diminishes.

When all of the water has been removed from the free water zone by the combination of the drains and some water draining into the base, *all movement of water into the drains ceases.* This is in spite of the fact that the capillary fringe is still saturated.

All further drainage after this point has been reached is determined by the drainage rate of the base. Gradually more and more water is pulled down into the base, and more of the topsoil above approaches field capacity.

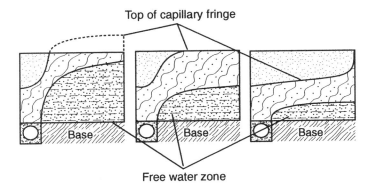

Figure 5.4. Shows the free water zone being lowered as water flows into a drain, and the capillary fringe following parallel to the top of the free water zone.

When a permeable topsoil is placed over a slower-draining base and subsoil drains are installed in that base, the rate at which the saturated free water zone of the topsoil will be diminished (drained) can be calculated using Hooghoudt's formula (Adams and Gibbs, 1994), which was developed for the agricultural sector. (See Figure 5.5.)

This formula calculates the rate at which the saturated free water zone of topsoil drains at the midpoint between the two drains. This is, of course, the slowest draining point, so this formula always gives the slowest rate of drainage over the total soil surface under consideration.

Remember that the capillary fringe always exists above this saturated free water zone in these types of profiles.

Hooghoudt's Formula

This formula is used by engineers and others to calculate the drainage rate between subsoil drains. The formula is as follows:

$$D = \frac{4 \times K \times H^2}{S^2} \text{ mm/hr} \qquad (A)$$

This formula is classically used for drains that are placed much deeper under the ground, but it can be adapted for other situations.

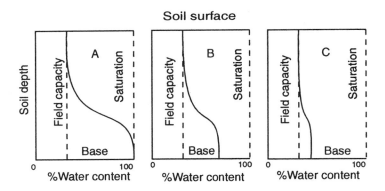

Figure 5.5. Water being removed by the base and the profile progressively approaching field capacity. In A a capillary fringe is present and the soil at the base is saturated. In B and C the whole profile is progressively approaching field capacity.

(See Figure 5.6.) To be able to use the formula properly in a range of situations it is useful to understand how it is derived. In the formula

$$D = \frac{4 \times K \times H^2}{S^2}, \text{ inches (or mm)/hr}$$

D is the drainage rate, in **inches** (or **mm)/hr**, of the saturated free water zone at the midpoint between the drains

K is the saturated hydraulic conductivity of the topsoil in **inches** (or **mm)/hr**;

H² is equivalent to $H_A \times H_B$ where:

  H$_A$ is the height of the saturated free water zone in the soil (as explained in the example above) in **inches, feet,** or **metres.** This is the component that relates to the pressure that is being exerted by the head of water above it in the soil. This value may be greater than the depth of the saturated free water zone, as it also includes the height of the head of water, which increases as the slope of the base increases. (See Figure 8.6.)

  H$_B$ is the height of the saturated free water zone, which the water has to pass through to move laterally, in **inches, feet,** or **metres.** This is the height that is a component of the cross-sectional area through which

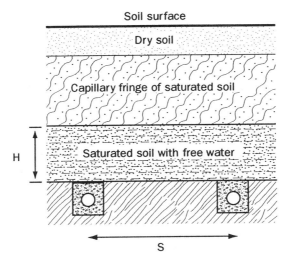

Figure 5.6. The depth H of saturated topsoil with free water, and the spacing S of subsoil drains.

the water has to flow, and it is the component that takes into account the constraint of flow outlined in Figure 4.7.

S^2 is equivalent to $S_A \times S_B$, where:

S_A is the distance between the drains in **inches, feet,** or **metres.** This is the distance which, when divided into H_A, gives the gradient on which the pressure of the head is calculated.

S_B is the distance between the drains in **inches, feet,** or **metres.** This is a component of the area over which the water was collected, as explained in Figure 4.7. (If metric measurements are used, i.e., the drain spacings are expressed in metres, then this figure is multiplied by one metre to give the cross-sectional area of the soil surface through which the water entered the profile).

Thus the formula now reads:

$$D = 4 \times K \times \frac{H_A}{S_A} \times \frac{H_B}{S_B}$$

So for most cases where there is very little gradient between the drains, H_A and H_B will be the same, and when multiplied together will be $\mathbf{H_A \times H_B = H^2}$.

In the same way, $\mathbf{S_A \times S_B = S^2}$.

Why is the 4 in the formula?

The distance the water has to move from the midpoint to the drain is not S, but rather one-half S, so the bottom of the formula should read

$$\frac{1}{\frac{1}{2}S_A \times \frac{1}{2}S_B} = \frac{1}{\dfrac{S_A}{2} \times \dfrac{S_B}{2}} = \frac{4}{S_A \times S_B} = \frac{4}{S^2}$$

So when the formula is being used to calculate the drainage rate of a soil between two drainpipes where there is little slope, formula A, as expressed above, can be used.

If there is only one drainpipe on the edge of a flower or shrub bed (as shown in Figure 8.6), then the 4 can be ignored in the formula, because the water has to travel all the way across the bed to reach the drain.

When the formula is used to calculate the rate of drainage for a saturated free water zone depth (H), and a drain spacing (S), this will only give a starting drainage rate. As the water drains down out of the top of the free water zone and is removed by the drains, the height of the saturated free water zone of soil is decreased, i.e., H diminishes.

This causes the drainage rate to decrease rapidly because as the height of the free water is decreasing, the H component in the formula is decreasing by the square of H (H^2).

For example, if the depth of the saturated free water zone of topsoil was 8 in. in the initial calculation, then 8 in. $\times$ 8 in. $= 64$ in^2. If the free water zone has been reduced to 6 in., then 6 in. $\times$ 6 in. $= 36$ in.^{2} (200 mm $\times$ 200 mm $= 40,000$ mm^2 and 150 mm $\times$ 150 mm $= 22,500$ mm^2). This means that the drainage rate will have been slowed down by about 45%, simply by the top 2 in. (50 mm) of the free water zone being depleted.

The drainage rate of a free water zone of topsoil sitting on a slow-draining base decreases rapidly as it drains, and as the height of the free water zone becomes smaller, e.g., when this height is only 2 in. (50 mm) deep, the rate at which water moves sideways toward the drain is extremely slow.

The relationship between the spacing of drains, the height of the free water zone in the topsoil, and the drainage rate is illustrated in Figure 5.7. It can be seen that gradient A, from the midpoint on the closer drain spacing at 16.5 ft (5 m), was steeper than gradient D for the same free water depth, but at the wider drain spacing of 33 ft (10 m).

Since the gradient reflects the component of gravity involved, it means that there will be a stronger "pull" by gravity in A than in D. Hence for the same free water depth, the water will move faster along gradient A than D, simply reinforcing the point that the closer the drains, the faster a profile will drain.

It also highlights the fact that as the free water depth decreases, so too does the gradient. Consequently, the rate of drainage decreases.

Figure 5.7 clearly illustrates the point that if the free water depth is shallower, drains have to be spaced much closer together.

This also applies to the total depth of topsoil. If the topsoil is shallow, e.g., 4 in. (100 mm), drains will have to be spaced very closely together to get any effective drainage rate. If the topsoil depth is much deeper, e.g., 12 in. (300 mm), the drains can be spaced much further

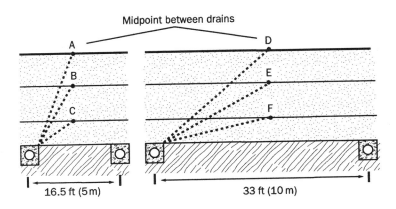

Figure 5.7. Gradients A, B, and C from the centre point between drains spaced at 16.5 ft (5 m) apart, for three different heights of saturated free water zones in a topsoil. Gradients D, E, and F are for the same heights of free water in the same topsoil, but with the drains spaced at 33 ft (10 m) apart.

apart to achieve the same drainage rate, simply because there is greater storage space in the profile for the free water zone.

Because both S and H are squared in the formula, as they vary they will change the drainage rate by the square of their distance apart or height, respectively, not just by their change in magnitude. In simple terms this means that if the drain spacings are halved, their effectiveness is fourfold. The same applies to the depth of topsoil. As a corollary to this, shallow topsoil and widely spaced drains will never work, other than to drain a very small area immediately adjacent to the drains.

In the United States, hydraulic conductivity is usually expressed as inches per hour, or sometimes as cm/hr. In many other parts of the world it is usually expressed as mm/hr.

How to Use Hooghoudt's Formula Using Nonmetric Units

This formula was derived in Europe and is usually shown in the literature in metric units. However, it is quite simple to use in imperial units provided a few simple rules are followed.

First, if the hydraulic conductivity (K) of the topsoil is expressed in in./hr, then the drainage rate (D) will be in in./hr. Similarly, if K is in cm/hr, then D will be in cm/hr.

Second, both H and S *must* be in the same units, as these will cancel each other out in the formula, leaving the result in the units of K. If the depth of the free water zone (H) is in inches, then the spacings between the drains (S) must be in inches. If H is expressed in feet, then S must be in feet.

The best way to demonstrate this is to work through a couple of examples.

Example 5.1. To calculate the drainage rate at the midpoint between drains spaced 10 ft (120 in.) apart in a soil with a drainage rate of 0.8 in./hr and when that soil has a free water zone of 0.75 ft (9 in.), the following calculation is made:

$$D \ (\text{in./hr}) = \frac{4 \times 0.8 \ (\text{in./hr}) \times (0.75 \ \text{ft})^2}{(10 \ \text{ft})^2} = 0.018 \ (\text{in./hr})$$

Note that both the drain spacings S and the depth of the free water H were in the same units—feet.

If we use the same data, but use inches instead of feet for H and S, then the calculation would be:

$$D \text{ (in./hr)} = \frac{4 \times 0.8 \text{ (in./hr)} \times (9 \text{ in.})^2}{(120 \text{ in.})^2} = 0.018 \text{ (in./hr)}$$

The result is the same. This demonstrates that it does not matter which units of measurement are used for H and S; the only thing that matters is that the units must be the same for both.

There is a very obvious question asked by most people—How do I know what is the depth of the free water zone?

This is very difficult to answer, but the calculations are very useful if you assume that the free water zone extends right up to the surface, as this will be the worst-case scenario. If this is done, do another calculation for the depth of the free water zone being lowered by 2 in. or by 3 in. This will give you an indication of how quickly water can be removed from the top of the profile. Usually it is very slowly if you are relying on the drains alone.

Note

Hooghoudt's formula becomes inaccurate when the drain spacing (S) approaches the same value as the height of the saturated free water zone (H). This is very rare in horticulture.

No doubt some aspects of the above may be difficult for some to understand in the first instance. However, the reader should not despair. In the next chapter we will work through a range of examples that will show how to use the formula even though the theory is not necessarily fully understood. In this regard, it is important that teachers lead students through the theory and demonstrate its application through tutorials and work on practical application in the field.

CHAPTER 6

Perched Water Table— The Concept and Its Use

Introduction

In the study of drainage the concept of a perched water table is one of the most difficult to understand; nevertheless, its understanding is essential for many aspects of sportsturf and horticulture.

The phenomenon occurs when a finer-textured material is placed over a coarser one, and the underlying coarse material has an influence over the drainage of water from the layer above. This may occur naturally where lenses of coarse sand or gravel are often covered by a layer of soil. It also occurs frequently in man-made landscapes and sportsgrounds where people have put down a layer of gravel or coarse sand and covered it with topsoil in the belief that it will drain. In reality this is frequently a complete disaster, because no water drains from the finer material until there is a capillary fringe built up above the coarser material.

In fact, there is usually a saturated zone of topsoil sitting above the coarser material, as described in the soil-filled tube example. (See Figure 5.2.) The only difference is that the height or depth of the capillary fringe is not as substantial over the gravel or coarser material as it is over an open-ended tube, as the gravel does exert some suction on the soil above.

However, if the height of the capillary fringe exceeds the depth of the topsoil above the interface with the coarse material, the topsoil will never drain downward.

A Perched Water Table in the Wrong Place

An example of this phenomenon observed by one of the authors involved a shrub bed where plants were obviously dying due to waterlogging of the topsoil. The bed consisted of the following construction: a base of heavy soil with a slope of about 1:60 toward one side: a 4-in. (100-mm)-deep layer of coarse sand over this base; and a subsoil drain cut into the bottom of the slope. There was about 10 in. (250 mm) of a good sandy loam over the coarse sand, the plants were planted in this soil, and there was a layer of mulch over the top.

You might well say, "What was wrong with that design? There was a good layer of topsoil, and a good drainage layer of clean coarse sand with a drain in it to remove the water as it drained from the topsoil. It all makes good sense."

Well, let us consider the problems with it. It simply did not drain, and when it was walked on, it was like walking on a water bed—it moved like jelly. The whole of the topsoil was saturated, and no water was draining down into the sand layer below. This was because a perched water table had been created, and the depth of the capillary fringe was theoretically higher than the depth of the topsoil. This meant that there probably needed to be another 8 in. (200 mm) of topsoil added to the top of the profile for any of the topsoil to have been able to drain back to field capacity.

In another example, one of the authors observed playing fields in Melbourne that had been built by laying a 4 in. (100 mm) layer of Scoria (a porous expanded volcanic material often used as a drainage filter material), then covering it with 10 in. (250 mm) of a sandy loam. After three weeks of dry weather the surface of the field was still wet from previous rain and irrigation events. A perched water table had been created that had a capillary fringe which extended above the depth of the topsoil. In the winter the topsoil was always saturated, and it was seriously damaged and cut up by sporting use.

Confused? Well at the time, one author certainly was, and many others have been caught in the same way. This phenomenon seems to defy logic, but so often principles of soil physics appear to be confusing to some practitioners in the field. They do not appear to conform to common sense.

When one expects something to happen in an apparently logical way—like water flowing from a sandy loam down into a clean sand below—it doesn't. It is logical, however, once one has the appropriate knowledge of soil physics.

In the next section we will endeavour to explain in lay terms how this phenomenon works, without using any complicated physics terms.

Revision

To refresh your memory we would suggest that you go back and thoroughly read the first several pages in Chapter 2 concerning how gravity, the forces of water adhesion to soil particles, and surface tension are relevant to drainage. In addition, read the first several pages in Chapter 5 regarding the concept of the capillary fringe.

An understanding of all of these concepts is needed to fully comprehend how this type of perched water table works.

How does the perched water table work?

Consider a tube filled with a soil (sand or gravel), with a coarse gauze covering placed over the bottom of the tube to prevent the soil from falling out. This tube is then filled with water and placed on a sink and allowed to drain freely until no more water drains out from the bottom. As explained in Chapter 5, a strange phenomenon occurs. One would expect gravity to pull all the water out of the column and all the soil in the tube to reach the same water content at around field capacity—after all there appears to be nothing preventing this from occurring. *Wrong!*

Water does drain out of the bottom of the column, and the soil at the top does reach field capacity, provided the column is high enough, but there will *always* be a zone at the bottom of the column that is saturated (or, for some purists, quasi-saturated). (See Figure 6.1.)

This zone is called a *perched water table.*

This saturated zone from the bottom of the tube up to the air entry point is also called the capillary fringe. In the tube the suction being applied at the base is zero, and there is no free water present at the bottom of the tube.

The top of the perched water table is the air entry point, above which there is sufficient pull by gravity to break the menisci of some of the large pores, which begin to drain. The water content decreases with height, until the field capacity is reached, at which point all water

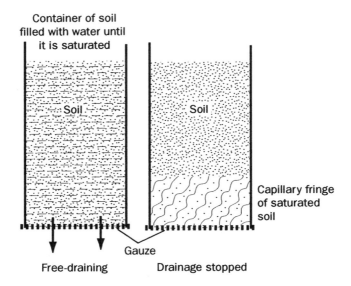

Figure 6.1. A column of soil that has had water added and been allowed to drain, and the eventual formation of the saturated zone or perched water table at the bottom of the tube.

movement stops and no further change in water content will occur as you go further up the tube.

The height of the perched water table is determined by the pore size distribution of the soil. In turn, this is related to the size of particles and how densely they are packed together. A fine-textured soil will have a deep saturated zone or deep perched water table, and a coarse gravel will have a shallow one.

For example, using the above method, a USGA sand with an average particle size of 0.4 mm may have a perched water table of about 7 in. (180 mm), but this can vary from about 5 to 9 in. (127 to 230 mm) in various sands that meet the USGA specification. A coarse washed river sand with an average particle size of 1.5 mm may have a perched water table of about 2 in. (50 mm), and a gravel with an average particle size of 4 mm may have a perched water table of about 0.75 in. (19 mm). The explanation for this is as follows:

When water is in a pore or between two surfaces, the "ends" of the water in the pore have menisci, which are the boundaries of the water with the air. All readers would have observed this on the top of a glass of water. (See earlier explanation for *surface tension* in Chapter 2.)

A similar sort of thing happens at the bottom of the soil column where it is exposed to air at the bottom of the tube. A drop forming at the bottom of the column will stretch the neighbouring menisci, and at some point these get stretched just enough to prevent a drop from falling. At this point water stops draining from the bottom of the tube.

The menisci at the top of the perched water table are stretched more than those at the bottom, because they have to carry the full weight of water in the column below them. Because the force of gravity cannot break these menisci in the large pores, no water will flow out of them. The combined force of surface tension in all the pores in the soil column will support much more water at the bottom of the column (perched water table) than it does when it is at field capacity.

Higher up in the soil column above the air entry point, water is able to move downward because it is in contact with more particles and pores from the soil below it. Water "flows" down these particles because it is being moved by the adhesive forces of the particles themselves, thus forming a vehicle for movement downward. When it reaches air at the bottom of the tube there is no vehicle on which the water can move, and the surface tension of the menisci will stop any further water movement downward. (See Figure 6.2.)

When more water is added to the top of the tube it will fill the larger pores in the soil above the perched water table. Gravity will then begin to pull some of this water through the profile and out the base. Some of the menisci in the large pores close to the surface break, and the extra weight of water above the base allows drops to form at the bottom of the column. If the water in the tube reaches an equilibrium just prior to more water being added to the top, i.e., there is no more flow out the bottom, then the amount of water coming out the bottom will be the same as that added to the top—the "one drop in, one drop out" situation.

Drops form at the bottom menisci, and this will continue to occur until the original equilibrium has been reached and the perched water table has been restored to its original equilibrium height. (See Figure 6.3.)

The height of a perched water table can be considered, for all intents and purposes, as constant; i.e., when water is added at the top

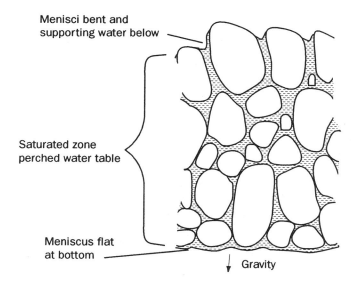

Menisci bent and
supporting water below

Saturated zone
perched water table

Meniscus flat
at bottom

Gravity

Figure 6.2. Water is held in the pores by the surface tension of menisci at the top and bottom of the column. The combined effect of surface tension and the adhesive forces of the soil particles is stronger than the force exerted by gravity. This prevents water from draining out the bottom.

it is temporarily raised, then as drainage ceases it quickly comes back to its original height.

The amount of water held in pores of different size categories is reflected by the force (or suction) needed to remove increasing amounts of water from soil after it has been saturated. This concept, i.e., the amount of water that is left in a soil after it has drained at a certain suction (or height above a free water table) is measured in the laboratory and expressed as the moisture release curve. (See Chapter 2.) Different soils will have different shaped moisture release curves, with different heights of perched water tables.

The height of the perched water table for a particular soil can be read from its moisture release curve. The shape and characteristics of this curve are reflections of the particle size distribution of a soil and the degree of compaction of the soil. Certain things can be predicted about the behaviour of the soil from these curves.

For USGA sands we prepare two curves in the laboratory for each sand. The first is generated when it has had only light compaction,

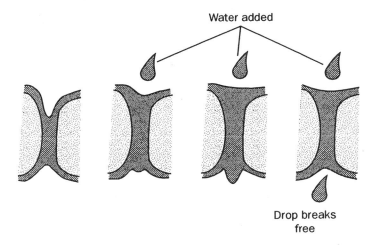

Figure 6.3. The meniscus bends and a drop forms, which breaks off and falls when water is added at the top.

and the second with quite heavy compaction, roughly equivalent to that expected by its use in the field.

If the two curves are very divergent, this means that a particular sand will compact under use, and with different levels of compaction across a sportsfield or golf green, some areas of that ground or green will behave differently than others. One area may be quite dry on the surface whereas another more highly compacted area will be quite moist, making grass and water management difficult. Sands with these characteristics should not be used to construct these types of facilities. (See Table 6.1.)

To illustrate this, four sands were chosen; three meet the USGA specification, but the fourth (sand D) has too many fines. From the particle size analyses (gradings) of the four sands, and from their two moisture release curves, it can be seen that each sand will behave quite differently. (See Figure 6.4.)

Sand A is extremely uniform and not affected by compaction, as shown by the fact that the two curves are almost identical. The perched water table is about 10 in. (250 mm) deep; its field capacity is about 4%; and if this soil were laid too deep, the top would be extremely droughty because the top of the profile would only hold 4% water only minutes after water was applied. This sand has been

Table 6.1. Shows four different sands and their particle size distribution as percentage by weight. Sands A, B, and C have been used to build leading stadiums in Australia, and Sand D is one that does not meet the USGA specification because it has too many fines in the 0.10 to 0.25 mm range.

Particle Size (mm)	Sand A (% by wt)	Sand B (% by wt)	Sand C (% by wt)	Sand D (% by wt)
> 2.0	0	0	0	0
1.0–2.0	0	1.4	1.6	0.4
0.5–1.0	3.0	23.5	24.0	15.9
0.25–0.5	93.0	60.4	49.4	40.4
0.10–0.25	4.0	12.4	15.5	41.4
< 0.10	0	2.3	9.5	1.9

used in Parramatta stadium, one of Australia's best-performing football stadiums.

Sand B is an excellent sand and was used in the reconstruction of QEII Stadium in Brisbane. It does compact a little, but it has a field capacity of about 8%, saturation occurs at about 32%, and it has a perched water table at about 8 in. (200 mm). There is little divergence in the two curves, which indicates that compaction will not adversely affect its surface moisture levels.

Sand C was used in Bruce Stadium in Canberra in 1991, and one can see that the shape of the curve is steeper than B because there are more fines. The depth at which this soil is laid is less critical than for A or B because it takes more depth for field capacity to be reached. The two curves are somewhat divergent, and this was reflected in the performance of the sand, as it did compact in the high-wear areas and the moisture content did differ at various locations over the field. Bruce Stadium was reconstructed in December 1997 using a much better sand, which was much closer to Sand B.

Sand D has a much higher fraction in the 0.10 to 0.25 mm range (41.1%). As a result of this, the perched water table is at about 12 in. (300 mm). This type of sand is too fine for perched water table construction because there are often problems of poor root growth due to too many fine pores. The two curves are quite divergent and indicate that this sand would behave differently under different levels of compaction.

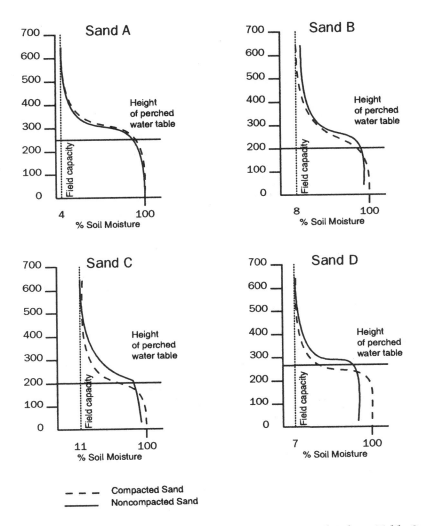

Figure 6.4. Moisture release curves for sands A, B, C, and D from Table 6.1, and the effect of compaction on the shape of the curve.

The reader can see that these four sands all have different shaped moisture release curves, influenced by the amount of fines in each sand. As a generalisation, the more fines, the deeper the perched water table.

Another point worth noting is the slope of the curve between the air entry point at the top of the perched water table and field capacity.

The narrower the range of particles, the flatter this section of the curve. In Sand A, this is somewhat flat because it is an extremely uniform sand, while in Sand C this part of the curve is fairly steep because this sand has a relatively wide range of particle sizes.

Once this concept has been understood, we can move on to consider what happens when we place fine soils or sands on top of coarser materials. In reality this is just what should happen when we construct a golf green, a bowling green, or a football stadium with a perched water table using these sands.

Usually these types of sands are placed on top of a much coarser sand or gravel, which covers all of a consolidated base. Drainage pipes are installed in trenches cut into the base, and these pipes are surrounded by gravel.

The presence of the perched water table in these profiles allows the use of a high-draining sand, which compacts only slightly under use. This sand is very good for rapid drainage, but if it were used by itself it would drain so quickly, and would hold so little water, that you could never grow high-quality grass on it. It would be too droughty in the summer.

The soil physicists have come to our rescue—allowing us to use these high-draining, low-compacting sands—by creating a perched water table profile. When the sands are placed over the correct gravel and the correct depth of top sand is used, the perched water table provides a reservoir of water for the grass to use, but the complete profile still drains very well. The authors strongly believe that not all profiles should have 12 in. (300 mm) of sand, as specified by the USGA (*USGA Record,* 1993).

When such a construction is made, the phenomenon of a perched water table comes into play. It also occurs in the bottom of plant pots, whether there is a drainage layer placed in the bottom of the pot or not. There will also be a perched water table in the bottom of planter boxes, and this is often why plants do so badly in some of these situations.

When sand, or soil, is placed directly on top of gravel or a coarser sand, a perched water table is formed in the finer material directly above the interface between the two materials. A saturated zone will extend from the interface with the gravel up to the air entry point. Now if the topsoil (or sand) depth is not as high as the theoretical depth of the perched water table (capillary fringe), then the whole of the soil will remain saturated and never drain.

When building these profiles (in a golf green, a bowling green, or a sportsfield) or filling a shallow pot or a planter box, one must always be aware of the height of the perched water table. If this is not understood and established correctly, the profile could be constantly wet or droughty on the surface, and plants will always be unthrifty or, even worse, may die.

The coarser the underlying gravel or sand, the deeper the perched water table will be. Conversely, the closer the average particle size of the underlying material is to that of the sand or soil above it, the shallower the perched water table will become. This is because when it rains, the sand above the perched water table becomes saturated, and as long as this situation persists, water is fairly quickly and easily drawn downward by gravity through the sand and into the underlying gravel. The gravel can conduct water much faster than the sand and never becomes saturated, provided there is an adequate system of drains to remove this water.

When the rain stops and water ceases to enter from the top, the water from the larger pores in the upper part of the profile drains out very quickly and this part of the profile rapidly approaches field capacity. Water is held in ever increasingly large pores by the menisci, until at the air entry point all the large pores are full (see Figure 6.5) and the perched water table reaches a constant height. From that point onward, only minute amounts of water drain out of the sand down into the gravel.

How does water move from the sand (or soil) down into the gravel?

For the sake of this discussion we will use a sand over a coarser gravel to illustrate the principles involved. At the interface between the two materials there will be a large number of pores in the sand, and only a few in the gravel. In other words there will be a large number of pores and particles in the sand which will not be touching any particles in the gravel below. These pores will be in contact with air. (See Figure 6.6.)

A small number of pores in the sand meet the surface of a gravel as a wall; the remainder contact air-filled spaces. This means that there is only a fraction of the sand particles that is actually in contact with gravel particles. This is crucial in the understanding of the whole process.

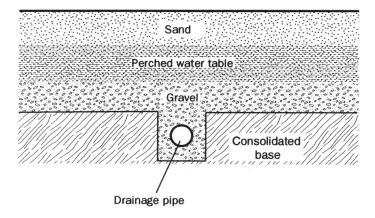

Figure 6.5. A typical perched water table situation occurring when a USGA sand is placed over a gravel, as occurs in golf and bowling greens.

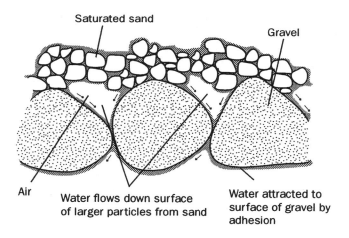

Figure 6.6. Water moves down from the fine sand above into the gravel below by flowing around the larger particles where they have contact with the sand above.

When there is heavy rain, or a lot of water reaching the interface from above, the water mainly enters the gravel from the large pores in the sand, as the menisci in these pores at the interface cannot hold the water back, and it flows, or drips, down from the spaces between the sand particles. The surface tension in these large pores is being broken by the weight of water above. Under these conditions the majority of the water enters the gravel by this means.

If the upper part of the perched water table has been raised by the extra water from above, the rate at which the majority of this water drains out into the gravel is at the saturated hydraulic conductivity of the sand. This will always be slower than the gravel's ability to accept it and remove it downward, provided there is somewhere for this water to be stored or otherwise removed by drains.

When the input of water from above ceases, the perched water table then begins to drain back to its equilibrium state, and the rate of downward movement of water slows quickly. This is because most of the large pores in the top of the profile have drained. It is then increasingly difficult for gravity to drain water down from the narrower pores against the combined forces of surface tension of the menisci in the smaller pores and the adhesive forces holding the water to the particles.

As flow slows to a point where the menisci in the large pores at the top of the perched water table are beginning to bend (as shown in Figure 6.2), all further flow from the bottom of the sand layer into the gravel will occur at the contact points between the sand and gravel particles.

The water continues to flow down the gravel particles because it is being pulled down by gravity, and also by the adhesive forces on the gravel particles.

The combination of these two forces pulls water from the pore spaces adjacent to the sand particles in touch with the gravel particles. This "flow" along the surface of the touching particles will continue until the combined forces of surface tension in the menisci of the pores in the sand and the adhesive forces of the sand particles in contact with the gravel equal those forces trying to pull the water down.

At this point an equilibrium will have been reached, and there will be a saturated perched water table above the interface. No further water moves downward out of the perched water table.

The Effect of the Gravel

When the sand is drained in the tube in Figure 6.1, the height of the perched water table will be determined by the average pore size of the soil or sand. It will also be as high as it can be because there is no suction at the bottom of the tube or, in other words, there are no particles in contact with the sand that will conduct water downward.

When there is gravel under the sand, the height of the perched water table is reduced as some of the water from the perched water table is drawn down, as described previously.

The extent to which the height of the perched water table is reduced is determined by the number of gravel particles that are in contact with the sand above. If the gravel is fine, or has many fine particles, there will be much more contact with the sand above and more water will flow down across these contacts.

The shape of the gravel particles is also important because if the gravel is a rounded river gravel, there will be only a small point of contact with the sand at the top of each pebble. If the gravel particles are flat and narrow, then when they pack together some of them will lie horizontally, thus presenting a much bigger surface area in contact with the sand.

Any two gravels may have a very similar sieve analysis, yet they can have a very different influence in lowering the height of a perched water table. For example, a rounded 3- to 5-mm gravel had a suction of 1.4 in. (35 mm), whereas an angular sharp 3- to 5-mm gravel had a suction of 2.8 in. (70 mm).

In lay terms this means that when a good USGA sand was placed over these two materials, the depth of the perched water table was 6.7 in. (170 mm) in the profile with the rounded material and 5.3 in. (135 mm) with the sharp gravel. These differences are very important because if the sand depth were 12 in. (300 mm) in the above situations, plant roots would have to be at least 5.3 in. (134 mm) deep on the rounded gravel, or 6.7 in. (170 mm) deep in the sharp gravel, just to reach the water in the perched water table. This is clearly far too deep; the roots should never have to reach more than about 4 in. (100 mm) from the surface of the top of the perched water table.

This is of particular importance in golf and bowling greens where the grass is cut very short. To expect good root systems to be constantly more than 6 in. (150 mm) deep—particularly in the middle of a hot summer—is a very tall order. If the roots do not reach the perched water table the grass will have to get its water from the sand above it, so the profile will perform—in relation to providing water to the grass—as if there were no perched water table present.

There is another problem when the roots do not reach the perched water table: nutrients are stored in this large volume of water, so if the roots cannot access it you have wasted money on fertilisers that will just be washed out the bottom of the profile. The only benefit of the perched water table in such greens is that they will still drain very quickly during rain.

This principle is essential when considering design and construction of golf and bowling greens and sand-based sportsfields.

A simple demonstration of the effect of the gravel and the number of particles touching the gravel can be set up using the following method. Take a laboratory sieve with a mesh size of 0.5 mm or smaller. Fill it up with sand and completely saturate the sand. Allow it to drain on a sink until all drainage ceases from the bottom of the sieve. Then lift the sieve up, still keeping it parallel to the floor, and place a finger onto the bottom of the mesh. Water will immediately begin to flow down your finger, and then it will stop. If you then place several more fingers in contact with the mesh, water will begin to flow down all of them again. This is what is happening with the gravel particles touching the sand above at the interface.

Because there will be much more of the sand in contact with air than with the surface of the gravel below, water will only flow downward along surfaces. Hence, most of the water in the sand will be "suspended" at the interface by the menisci in the sand layer. (See Figure 6.2.) *This is why a perched water table exists.*

It should be noted that there will be an equilibrium reached between the two layers, and the more uniform in particle size each of the two materials are, the more quickly this equilibrium is reached. *There will always be a perched water table in the sand above the gravel.* This will occur right up until the two materials are so similar that they contain the same amount of water at field capacity.

The behaviour of soils (as distinct from sands) in relation to the formation of perched water tables can not usually be predicted from their particle size analysis alone. Other measurements, and a deeper understanding of other properties—such as the percentage of organic material and the presence or absence of clay aggregates—are required to understand how it is likely to react during and after construction.

The perched water table in soils is affected to a much greater degree by compaction than sands. When soils are compacted the perched water table (capillary fringe) increases in height and can often extend above the surface of the soil in these compacted areas. This is of great importance in situations where soil is used in perched water table constructions, such as planter boxes.

Capillary Fringe (see also Chapter 5)

The perched water table in the above situation is a capillary fringe. This is very important to understand, because when one digs a hole in it, no water will flow into the hole. This is why a golf hole can be placed into the saturated perched water zone of a green and no water will flow into the hole. It is also why, when a trench is dug right down to the gravel, there is no free water seen in this trench. (See Figure 6.3.)

Perched Water Table on a Slow-Draining Base

There is another situation that is often referred to as a perched water table. This is the "free water" zone, which is dealt with in Chapters 5 and 7.

The free water zone situation occurs when there is an impermeable base, or a very slow-draining base, with an overlying layer of permeable topsoil. The saturated free water zone that forms at the top of the base is sometimes called a perched water table. (See Figure 6.7.)

This differs from the situation explained above because it has a coarser material over a finer one. Its presence is usually temporary: as the base slowly drains, the height of this perched water table diminishes. In the perched water table cases considered above, the perched

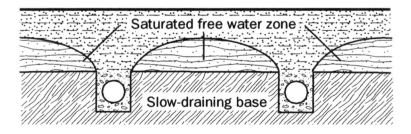

Figure 6.7. A perched water table of free water sitting on a slow-draining base between two subsoil drains.

water table always exists, and the only way water can be removed is by evaporation or by transpiration from grass or plant growth.

Remember, there will always be a capillary fringe over this perched water table above the slow-draining base. The saturated free water zone that exists for long periods of time between drains is also called a perched water table by some. In this instance it will be slowly diminished by both the drainage of the base and the drains.

The characteristic of this type of perched water table is that it is temporary, as there is always some suction being applied to it by the base. This is, of course, unless the base is rock or some other material that is completely impermeable.

CHAPTER 7

How to Calculate Drain Spacings for Various Soils

Chapters 4, 5, and 6 explained how water moved laterally in soil over a slow-draining base and how Hooghoudt's formula was derived. This formula can be used to calculate the drainage rate of many of the situations that occur on golf courses, football fields, racecourses, shrub beds, or other sportsturf or horticultural situations where subsoil drains are already installed, or are intended to be installed.

On most soil profiles that enable a good grass sward to be grown, excluding sand-based perched water table profiles, some estimates of the hydraulic conductivity of the soil can be made that will allow you to do calculations on the expected drainage rate of particular soils.

These can be used to make an estimate of how far apart subsoil drains would have to be spaced to effectively drain the surface of the soil at the midpoint between the drainpipes.

The dispersion test outlined in Chapter 1 is worth doing here. A piece of soil is placed into distilled water, and if the water immediately becomes cloudy the soil is unstable. This would also indicate that the soil would have a low hydraulic conductivity under compaction.

However, there is only one sure way to find out the actual hydraulic conductivity of a particular soil, and that is to have it tested by a laboratory which has this test available. The test outlined in Chapter 12 is an excellent one to determine hydraulic conductivity under different levels of compaction that might be experienced in the field. If this test is not available, then use whatever data are available, but remember that whatever test is carried out by a laboratory, it should be at several different compaction levels to give a better understanding of the performance of the soil in the field situation.

There are several laboratories in Australia that now carry out tests to provide clients with information on their soil that allows judgments to be made on its performance in the field. Engineers are now writing contract supply specifications around these tests. These include the particle size analysis (the least important), the water holding capacity at 1 m suction, the compacted hydraulic conductivity at several levels of compaction, and bulk densities at these compaction levels. A typical soil specification may state that the soil must have a saturated hydraulic conductivity in excess of 0.4 in. (10 mm)/hr at 16 drops, a water holding capacity at 1 m suction in excess of 12 % by weight, and it may specify that the bulk density at 16 drops must not exceed 1.6 g/cm^3.

The estimates in Table 7.1 may help in any calculations. The hydraulic conductivity of these types of soils can vary greatly. It is mainly influenced by the amount of fines in the soil, in particular the amount of silt, as this is the most mobile of soil particles. Clay particles, which are much smaller than silt, usually aggregate and behave as much larger particles.

In many soils the silt/clay ratio is an important indicator of its potential behaviour. If this ratio exceeds 2, i.e., twice as much silt as clay, and the total percentage of silt and clay present is between 25 % and 35 %, the soil will very likely behave in an unstable manner. This means it will probably compact badly under use, or when it is worked while wet.

If the soil has a wide range of different sized particles it will compact, and this will also slow down the hydraulic conductivity.

It is extremely rare for any soil in a sportsfield situation to have a hydraulic conductivity of more than 2 in. (50 mm) per hour, unless it is a USGA sand. In these cases it will have to have a perched water table structure for it to be manageable and to grow grass in the summertime.

In the past there has been a common engineering practice that involved the installation of subsoil drains at 10-m centres in ordinary soil profiles that were often only 4 in. (100 mm), or at best 8 in. (200 mm) deep. There are very few fairways, football fields, etc., that have soil profiles deeper than 8 in. (200 mm), and in many cases they are as shallow as 4 in. (100 mm).

Table 7.1. Descriptions of various soils, and an estimate of their hydraulic conductivities in the field.

Soil Type	Soil Description	Estimate of Hydraulic Conductivity (in./hr and mm/hr)
Silt/clay loam 20–45% fines, i.e., particles less than 0.1 mm	Heavier soil that does not feel sandy, and when moist, can be rolled between the hands into a tube that does not easily break. These soils are common on golf fairways, football fields, school grounds, parks, etc.	0.2–0.4 in./hr 5–10 mm/hr
Sandy loam < 20% fines, i.e., particles less than 0.1 mm	Soils that still have fines in them but are much sandier. These soils will fall apart when an attempt is made to roll a moist sample between the hands. *Some* of these soils will drain faster than the above soils, but some will still compact and have low hydraulic conductivities.	0.4–2.0 in./hr 10–50 mm/hr

It is probably fair to assume that the top 2 in. (50 mm) of any soil profile that is being used for sport has to be drained to about field capacity to allow for reasonable use and playability. This means that the free water zone must be drained to allow the base to draw the capillary fringe down so this top 2 in. (50 mm) can be drained back to field capacity and mechanical stability.

Let us then consider how long it takes for the free water zone to be drawn down by 2 in. (50 mm) if it is relying on drains alone, i.e., assuming there is no water being removed down through the base. The following calculations will estimate how long it will take for the *midpoint between the drains* to reduce the height of the free water zone on three different soil types and at two different soil depths.

Hooghoudt's formula as outlined in Chapter 5 has been used to make the calculations in the following examples and in Tables 7.2, 7.3, and 7.4.

The formula is as follows:

$$D = \frac{4 \times K \times H^2}{S^2}$$

D = drainage rate of the top of the free water zone at the midpoint between the drains, in inches per hr or millimetres per hour.
K = hydraulic conductivity of the soil, inches per hour or millimetres per hour.
H = the depth of the saturated free water zone of topsoil, in inches, feet, or metres.
S = the spacings between the drains, in inches, feet, or metres, remembering that the units used for H must be the same as those used for S.

Example 7.1. Let us assume that the whole 8 in. (0.2 m) depth of topsoil is saturated and the free water zone extends to the surface, as would occur after prolonged rain on a sandy loam soil with hydraulic conductivity of 0.8 in. (20 mm)/hr, and drains spaced at 33 ft, or 400 in. (10 m) apart.

$$D = \frac{4 \times 0.8 \times 8 \times 8}{400 \times 400} = 0.0013 \text{ in. } (0.032 \text{ mm})/\text{hr}$$

This rate is only the starting rate, and will slow down as the height of the saturated free water zone diminishes. When this zone has been lowered by 2 in. (50 mm), i.e., the height of the free water zone will be 6 in. (0.15 m), the drainage rate will be:

$$D = \frac{4 \times 0.8 \times 6 \times 6}{400 \times 400} = 0.0007 \text{ in. } (0.018 \text{ mm})/\text{hr}$$

The average drainage rate for the top 2 in. (50 mm) will be:

Table 7.2. Drainage rate of the top 2 in. (50 mm) of a soil profile with free water zones of 4 in. (100 mm) and 8 in. (200 mm) depth; a hydraulic conductivity of 0.2 in. (5 mm)/hr; an air-filled porosity of 15%; and drains spaced at 3.3, 6.6, and 33 ft (1, 2, and 10 m) apart.

	Soil with a Hydraulic Conductivity of 0.2 in. (5 mm)/hr				
	Depth of Saturated Free Water Zone of Topsoil				
	8 in. (200 mm)			4 in. (100 mm)	
Distance Drains Spaced Apart, ft (m)	Drainage Rate of Topsoil, in./hr (mm/hr)	Time Taken to Drain Top 2 in. (50 mm) of Topsoil, hr (days)	Drainage Rate of Topsoil, in./hr (mm/hr)	Time Taken to Drain Top 2 in. (50 mm) of Topsoil, hr (days)	
33 (10)	0.0002 (0.005)	1200 (50)	0.00005 (0.0013)	6000 (250)	
6.6 (2)	0.0063 (0.16)	48 (2)	0.0012 (0.031)	240 (10)	
3.3 (1)	0.025 0.63	12 (0.5)	0.005 (0.125)	60 (2.5)	

$$\frac{0.0013+0.0007}{2}=0.001 \text{ in. } (0.025 \text{ mm})/\text{hr}$$

As can be seen, this is an extremely slow drainage rate. If the drains were spaced only 3.3 ft or 40 in. (1 m) apart, which increases the number of drains tenfold, then the rate would be:

$$\frac{4\times0.8\times8\times8}{40\times40}=0.128 \text{ in./hr}; \qquad \frac{4\times0.8\times6\times6}{40\times40}=0.072 \text{ in./hr}$$

$$\frac{0.128+0.072}{2}=0.1 \text{ in. } (2.5 \text{ mm})/\text{hr}$$

If the air-filled porosity of this soil were 20%, it would require 0.4 in. (10 mm) to be removed from the top 2 in. (50 mm) of topsoil for it

Table 7.3. The drainage rate of the top 2 in. (50 mm) of a soil profile with free water zones of 4 in. (100 mm) and 8 in. (200 mm) depth; a hydraulic conductivity of 0.8 in. (20 mm)/hr; an air-filled porosity of 20%; and drains spaced at 3.3, 6.6, and 33 ft (1, 2, and 10 m) apart.

Soil with a Hydraulic Conductivity of 0.8 in. (20 mm)/hr				
	Depth of Saturated Free Water Zone of Topsoil			
	8 in. (200 mm)		4 in. (100 mm)	
Distance Drains Spaced Apart, ft (m)	Drainage Rate of Topsoil, in./hr (mm/hr)	Time Taken to Drain Top 2 in. (50 mm) of Topsoil, hr (days)	Drainage Rate of Topsoil, in./hr (mm/hr)	Time Taken to Drain Top 2 in. (50 mm) of Topsoil, hr (days)
33 (10)	0.001 (0.025)	400 (17)	0.0002 (0.005)	2000 (83)
6.6 (2)	0.025 (0.63)	16 (0.67)	0.005 (0.125)	80 (3.3)
3.3 (1)	0.1 (2.5)	4 (0.17)	0.02 (0.5)	20 (0.83)

to reach field capacity—20% of 2 in. = 0.4 in. It would therefore take 5 hr for the top 2 in. (50 mm) of this profile to drain to field capacity, i.e.,

$$\frac{2 \text{ in.}}{0.4 \text{ in.}} = 5 = \frac{50 \text{ mm}}{10 \text{ mm}}$$

This means, of course, that the top 2 in. (50 mm) of topsoil will only drain if it is above the capillary fringe. In a great number of shallow profiles this simply does not happen, as the saturated capillary fringe will extend to near—or indeed above—the surface in some cases. This occurs particularly in compacted areas.

These same calculations have been carried out for soil types with different hydraulic conductivities, using three different drain spacings.

From Table 7.2 it is obvious that if the subsoil drains are spaced at 33 ft (10 m), they are all but useless in having any effect on the

Table 7.4. Shows the drainage rate of the top 2 in. (50 mm) of a soil profile with free water zones of 4 in. (100 mm) and 8 in. (200 mm) depth; a hydraulic conductivity of 2 in. (50 mm)/hr; an air-filled porosity of 25%; and drains spaced at 3.3, 6.6, and 33.3 ft (1, 2, and 10 m) apart.

	Soil with a Hydraulic Conductivity of 2 in. (50 mm)/hr				
	Depth of Saturated Free Water Zone of Topsoil				
	8 in. (200 mm)			4 in. (100 mm)	
Distance Drains Spaced Apart, ft (m)	Drainage Rate of Topsoil, in./hr (mm/hr)	Time Taken to Drain Top 2 in. (50 mm) of Topsoil, hr (days)		Drainage Rate of Topsoil, in./hr (mm/hr)	Time Taken to Drain Top 2 in. (50 mm) of Topsoil, hr (days)
33 (10)	0.0025 (0.063)	200 (8)		0.0005 (0.013)	1000 (42)
6.6 (2)	0.062 (1.563)	8 (0.33)		0.123 (3.13)	40 (1.6)
3.3 (1)	0.25 (6.25)	2 (0.08)		0.05 (1.25)	10 (0.42)

general drainage of the area in between the pipes. It would be a complete waste of time and money to install subsoil drains 33 ft (10 m) apart on a topsoil that had a profile depth of 8 in. (200 mm), let alone 4 in. (100 mm) deep.

Even with the drains spaced at 3.3 ft (1 m), it would take 12 hours to drain the top 2 in. (50 mm), midway between the two pipes, and two and a half days if the free water depth were 4 in. (100 mm). Remember that here we are still only talking about the depth of the free water zone, and not the depth of the topsoil.

Let us now consider the same set of calculations on a faster-draining sandier soil with a field hydraulic conductivity of 0.8 in. (20 mm)/hr.

From Table 7.3 it can be seen that it would take a very long time, 17 days, for the top 2 in. (50 mm) of a profile with a 8 in. (200 mm) free water zone to drain at the midpoint between the drains if they were spaced at 10 m, and the capillary fringe were below the soil surface.

A soil that drains 0.8 in. (20 mm)/hr is a very good-draining soil in the field, but installing drains at 33-ft (10-m) spacing is a complete waste of time and money.

Even with the drains spaced at 1 m, it would take 4 hr to reduce the free water zone by 2 in. (50 mm) at the midpoint between the drains.

If the drains were spaced at 6.6 ft (2 m) it would take 16 hr for the midpoint to be reduced by 2 in. (50 mm), and this is probably an acceptable drainage rate in some circumstances. The profile would have to be 1 ft (300 mm) deep of a good sandy loam, and be regularly decompacted. These sandier loams will have a lower capillary fringe because they have fewer fine pores than the heavier silt/clay loams.

The sandier higher-draining topsoils that are 10 to 12 in. (250 to 300 mm) deep will be able to drain the top 2 in. (50 mm) quickly, because the capillary fringe will rarely extend to the surface. However, these profiles with subsoil drains are very costly, and in most circumstances would be uneconomic.

Let us now consider a soil with a hydraulic conductivity of 2 in. (50 mm)/hr. This would be considered a very good-draining soil. Such a soil is very sandy, and in fact many bowling and golf greens would not drain at the rate of 2 in. (50 mm) per hour.

From Table 7.4 it can be seen that, if the free water zone is 8 in. (200 mm), drainage will still be very slow, even if the drains are spaced at 2 m. Drains spaced at 1 m will still take 2 hr to drain the top 2 in. (50 mm) at the midpoint between the drains, if it is above the capillary fringe.

A soil with a hydraulic conductivity of 2 in. (50 mm)/hr will have a much lower capillary fringe than the other soils used above. Hence the surface will drain a lot faster when water is being removed from the free water zone.

However, it is not a good practice to place these high-draining sandy soils straight over a slow-draining base because the soil receives water very quickly, which rapidly saturates the profile and then drains slowly. This very sandy type of profile quickly becomes droughty in the summer. It does not hold nutrients well in the root zone and can become waterlogged in the winter.

Rapid-draining sandy loams often are slower to drain and reach mechanical stability than heavier soil profiles during wet weather.

This is because they accept much more water from rainfall, as the surface of the heavier soils becomes saturated during rain, and then most of the subsequent rain runs off the surface. On the other hand, the sandy soil has a greater pore space, so it can accept much larger volumes of water, and a free water zone will be established more quickly during the same rainfall event.

Because of this higher permeability, and the larger storage space for water in the profile, there are often much larger volumes of water to be removed from the sandier soil. If the base is draining at the same rate as the heavier topsoil profile, it will take longer for the sandier profile to drain.

Unless subsoil drains are spaced very closely they will not significantly speed up this drainage across the whole profile.

On the sandier profile the area immediately above the drains and a small area on either side of them will drain very quickly. However, there can still be large areas of saturated soil with free water remaining in between the drains for long periods, waiting to be drained by the base.

There has to be a compromise in selecting an appropriate soil. On one hand, the soil needs to drain at a reasonable rate and should not fill up very quickly with water every time it rains. On the other hand, the soil needs to be able to hold enough water to grow good grass and not be too droughty in the summertime.

Rapid-draining sands should only be used in a perched water table (USGA) type profile.

In summary, what does all this tell us?

First, it confirms that water moves sideways in soil very slowly. Secondly, it highlights that in the past there has been a great deal of suspect thinking about how effective subsoil drains are in draining sporting facilities, particularly the traditional shallow topsoil profile laid over a heavy clay base.

We know there are some practitioners out there who might say that these figures are rubbish, because they have profiles with subsoil drains spaced at a particular spacing in a soil with a certain depth, and it drains much quicker than the figures in the tables.

In some instances they might be correct, but these will almost certainly be cases where the base is draining at a reasonable rate, and this rate is added on to the drainage rate achieved by the drains. In fact, in most situations the drains only remove a small portion of the water. The above calculations are a good estimate of the drainage rate, if it is only the drains that are removing the water.

For example, let us take a base that drains at the very slow rate of 0.01 in. (0.3 mm)/hr, and see how much effect this will have on the results as shown in Table 7.2.

It can be calculated that for a saturated free water zone of 8 in. (200 mm) profile with a soil having a hydraulic conductivity of 0.2 in. (5 mm)/hr, a base that drains at the very slow rate of 0.01 in. (0.3 mm)/hr would remove water from this zone (at the midpoint between the drains), at the same rate as subsoil drains spaced 4.6 ft (1.4 m) apart. It also drains a 4-in. (100-mm)-deep free water zone two and a half times faster than subsoil drains spaced 3.3 ft (1 m) apart.

If you compare the results shown in Table 7.5 with those of Table 7.2, it can be seen that the rate of drainage of the top 2 in. (50 mm) of the profile is influenced more by the drainage rate of the base than by the subsoil drains in all but one case, i.e., the 8-in. (200-mm)-deep free water zone with drains spaced at 3.3 ft (1 m).

The rate of 0.01 in. (0.3 mm)/hr is a very slow rate of drainage. Nevertheless, it still drains the free water faster than drains spaced 4.6 ft (1.4 m) apart on normal silt clay soils.

Given this knowledge, it becomes very clear that it is far more important to have a good-draining base than to install subsoil drains. It also explains why the process of vibra-moling—or any process that opens up the base and makes it drain better—works so well if done properly. Remember that the work has to be done while the subsoil is sufficiently dry for it to fracture or crumble. Any work that is done while the soil is wet enough to smear will have an adverse effect, in that it will make future drainage worse.

If during construction a base is graded or worked when it is too wet, it can easily be rendered impermeable, as was assumed in Tables 7.2, 7.3, and 7.4. This is also assumed in Figure 7.1.

Figure 7.1 shows how slowly a normal silt/clay soil drains after the whole profile has been saturated or a significant depth of free water has built up above the base. It can be seen that it takes 12 hr for the free water depth to be lowered by 2 in. (50 mm) in a strip of only 1 m wide, i.e., 20 in. (0.5 m) either side of the pipe.

Table 7.5. The drainage rate of the top 2 in. (50 mm) of a soil profile with free water zones of 4 in. (100 mm) and 8 in. (200 mm) depth; a hydraulic conductivity of 0.2 in. (5 mm)/hr; an air-filled porosity of 15%; and drains spaced at 3.3, 6.6, and 33.3 ft (1, 2, and 10 m) apart. It also takes into consideration the effect of a base that drains at 0.12 in. (0.3 mm)/hr.

	Soil with a Hydraulic Conductivity of 0.2 in. (5 mm)/hr and a Base that Drains at 0.01 in. (0.3 mm)/hr			
	Depth of Saturated Free Water Zone of Topsoil			
	8 in. (200 mm)		4 in. (100 mm)	
Distance Drains Spaced Apart, ft (m)	Drainage Rate of Topsoil, in./hr (mm/hr)	Time Taken to Drain Top 2 in. (50 mm) of Topsoil, hr (days)	Drainage Rate of Topsoil, in./hr (mm/hr)	Time Taken to Drain Top 2 in. (50 mm) of Topsoil, hr (days)
33 (10)	0.0002 + 0.012 = 0.0122 (0.006 + 0.3 = 0.306)	24.5 (1)	0.00005 + 0.012 = 0.01205 (0.003 + 0.3 = 0.303)	24.7 (1)
6.6 (2)	0.0063 + 0.012 = 0.0183 (0.16 + 0.3 = 0.46)	16.3 (0.67)	0.0012 + 0.012 = 0.0132 (0.031 + 0.3 = 0.331)	22.7 (1)
3.3 (1)	0.025 + 0.012 = 0.037 (0.63 + 0.3 = 0.93)	8 (0.33)	0.005 + 0.012 = 0.017 (0.125 + 0.3 = 0.425)	17.6 (0.73)

Note that only the area immediately adjacent to a pipe drains quickly, and for the free water to be reduced by 2 in. (50 mm) only 40 in. (1 m) away from the pipe will take 24 hr. Even after 96 hr the distance from the pipe that the height of the free water zone has been lowered by 2 in. (50 mm) will be less than 6.6 ft (2 m).

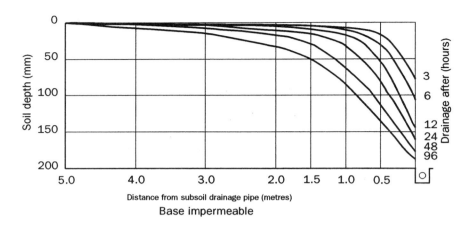

Figure 7.1. A soil with a hydraulic conductivity of 0.2 in. (5 mm)/hr, a saturated free water zone of 8 in. (200 mm), and a subsoil drain installed in the base. The amount drained after 3, 6, 12, 24, 48, and 96 hr is shown. The area under the curves is still in a saturated free water state. (From McIntyre and Jakobsen, 1998.)

In Figure 7.2 the same criteria as for Figure 7.1 were used, but the influence of the base draining at 0.012 in. (0.3 mm)/hr is added on. It can be seen that the drainage of the free water from the profile was much more effective, as the top 2 in. (50 mm) of the profile drains faster.

The comparison between Figures 7.1 and 7.2 clearly demonstrates the role of the base in the drainage of these types of profiles. These figures reinforce the great importance of having a base that drains at a reasonable rate, and clearly demonstrate how ineffective subsoil drains are by themselves.

The major lesson to be learned from this chapter is that subsoil drains do not remove large amounts of water from the profile, and any water removed is done so very slowly. Their installation is usually an insurance policy to guard against the base not draining, but this is only effective on profiles that are deeper than 10 in. (250 mm) and the drains have to be spaced at 6.6 ft (2 m) apart, or closer.

Subsoil drains only begin to become effective when the topsoil depth is more than 8 in. (200 mm). As the depth of topsoil increases, so too does the effectiveness of the drains, provided they are closely spaced.

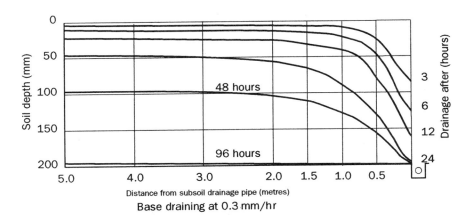

Figure 7.2. Shows a soil with a hydraulic conductivity of 0.2 in. (5 mm)/hr, a 8 in. (200 mm) saturated profile, a subsoil drain installed into the base, and a base that drains at 0.01 in. (0.3 mm)/hr. The amount drained after 3, 6, 12, 24, 48, and 96 hr is shown. The area under the curves is still in a saturated free water state. (From McIntyre and Jakobsen, 1998.)

Drains have to be spaced quite close together—as close as 3.3 ft (1 m)—to have any great effect in most soil profiles. The base will normally remove water quicker than drains spaced any further apart. The old engineering practice that we see all around the world where subsoil drains have been installed at 33.3-ft (10-m) intervals on sporting facilities leads to *a complete waste of time and money.* They would only ever drain an area about 3.3 ft (1 m) on either side of the drain after several days, provided the capillary fringe did not extend above the surface of the soil profile, which is likely on most shallow heavier soil profiles used for sporting facilities. Spacing the drains at even 6.6-ft (2-m) centres would have little useful effect on such profiles.

Subsoil drains that are fairly closely spaced are quite useful in removing ponded water from the surface. However, once this water disappears below the soil surface they become far less useful, as outlined above.

Subsoil drains work well in annual flower beds, as the soil generally has a high hydraulic conductivity, due to a high organic content. Usually a good depth of topsoil is used, often around 12 in. (300 mm), and the capillary fringe in these soils with high organic content is usually much lower than soils with little organic content.

Subsoil drains can ensure that shrub beds do not remain water-logged, and this is important for good growth. However, please remember they are rarely of much use on most sportsfields. So sit down, make some assumptions on the hydraulic conductivity of your soil, measure its depth, and then do some calculations and determine whether drains are really worth installing.

Notes

1. The type of drainage discussed in this chapter should not be confused with slit drainage, which is actually a type of surface drainage dealt with in Chapter 11.
2. The authors are aware that Hooghoudt's formula becomes inaccurate as the depth of profile approaches the distance between the drains. Adjustments have been made for this in Figures 7.1 and 7.2. For the calculations in the rest of the chapter, and the level of accuracy required for the practical turf manager, the formula works well.
3. The authors are also aware that there are formulas that determine the combined drainage rate of the topsoil and the base (Adams and Gibbs, 1994), but for the purposes of these shallow profiles the addition of the drainage rate onto the drainage rate of the topsoil gives a good estimate of what actually happens.

CHAPTER 8

Designing Subsoil Drainage Systems

As explained in Chapter 4, water moves laterally in soil very slowly. Water that falls on the surface of the soil takes a long time to reach subsoil drains. Hence it takes a long time for the top of the soil profile to drain to field capacity if it has to rely on this sideways movement of water to remove it through subsoil drains.

As a result of this concept being almost universally misunderstood by horticulturists, landscape architects, and sadly, many engineers, subsoil drains have been installed in all kinds of horticultural and sportsfield situations with little or no hope of their ever working properly. The expectations for performance of these drains has been far too high because assumptions about the extent that water would move sideways in the soil were far in excess of reality.

The complete lack of knowledge of the capillary fringe and its effect on the movement of water from shallow profiles has been another factor in the misunderstanding of how water gets into subsoil drains. In many cases, because of the shallowness of the profile and the presence of a capillary fringe that allows saturated soil to exist right up to the surface, drainage systems that have cost a lot of money to install could never have worked.

As a consequence, we have found sportsfields with 4 to 6 in. (100 to 150 mm) layers of topsoil with drains spaced at 33-ft (10-m) centres constructed all over the country, and for that matter all over the world. These drains could only ever drain the top 4 in. (100 mm) of this topsoil to a distance of about 12 in. (300 mm) from the pipe in any reasonable time. They could never drain the area in between the pipes. This approach is a complete waste of money.

Subsoil drains have been placed at the ends of slopes in large shrub and flower beds, and have only ever drained a tiny area of these beds.

The calculations in Chapter 7 give a graphic demonstration of just how slow these drains actually are in removing water—unless they are spaced very close together—in almost all situations. They only remove the water from the saturated free water zone, and not from the saturated capillary fringe above it.

There are only a limited number of situations where subsoil drains are going to be of much use.

Deciding Where to Install Subsoil Drains

When designing a new facility, or looking at a problem in an old one, the installation of subsoil drains can sometimes solve a drainage problem. Don't rush in and specify subsoil drains as a matter of routine. First, evaluate whether they will be of any real value in draining anything other than a small area directly adjacent to the pipe.

Sportsfields

Table 7.2 demonstrates that drains spaced at 33-ft (10-m) centres are completely useless for a sportsfield with 8 in. (200 mm) of topsoil, which has a compacted hydraulic conductivity of 0.2 in. (5 mm)/hr. Once the profile became saturated, it would take 50 days to drain the top 4 in. (100 mm) of the profile at a point only 16 ft (5 m) from the pipe, and the surface would probably still be saturated by the capillary fringe.

Surface Fall

The drains would have to be spaced at about 6.6 ft (2 m) for any reasonable drainage to occur. This would be very expensive, and would almost certainly never be incorporated in a large-scale facility. A much better strategy under these circumstances would be to make sure there is sufficient surface fall of about 1 in 70 (1 in 100 is definitely not sufficient), as this will shed a large amount of water before it enters the soil in heavy or prolonged rain. (See Chapter 10.)

Good Draining Base

Before spreading the topsoil, the base should be ripped to a minimum depth of 12 in. (300 mm), harrowed to break up large clods, and gypsum applied at a rate of 10 lb/100 ft^2 (500 g/m^2). This will ensure that the base will drain and there will be no areas that have been sealed by earthmoving equipment during the preparation of the base. It will maximise the effectiveness of drainage in the base.

The effect of a properly draining base on the drainage of the topsoil is clearly demonstrated in Figure 7.2. A very slow-draining base with a drainage rate of 0.012 in. (0.3 mm)/hr, will drain the top 2 in. (50 mm) of topsoil at the same rate as subsoil drains spaced 4.5 ft (1.4 m) apart. If the base drained at 0.04 in. (1 mm)/hr, it would be equivalent to drains spaced at 2.62 ft (0.8 m). It can be seen that a base that drains well is very much more important than having drains spaced at 3 ft, or less than 1 m, apart.

There is an even more important aspect of a draining base, as it has the advantage that it will drain the capillary fringe, whereas if the base is impermeable, or almost impermeable, once the drains have depleted the saturated free water zone, no water will move into the drains.

Clearly, from this, there are very few situations in sportsfields constructed with loam soil where there is any value in installing subsoil drains unless they are spaced very close together. By far the better and cheaper option is to ensure that the base drains properly and there is sufficient surface fall to shed unwanted water, as outlined above.

If confronted with an existing situation where the base is not draining effectively, it is far better to try to get the base to drain at a reasonable rate. This can be achieved by vibra-moling, or by using a verti-drain or some other compaction-breaking machine that shatters the base when it is dry. As mentioned previously, this work should never be carried out when the base is wet, otherwise it will probably make things worse.

Subsoil drains are of very little value in new sportsfields unless they are spaced at a minimum of 6.6 ft (2 m), and only then if the topsoil depth is 10 in. (250 mm) or more. A properly constructed draining base is far more important than installing subsoil drains.

Removal of Surface Water in a Saturated Profile

When heavy rain falls and the profile is saturated, there will often be free water remaining on the surface after surface drainage has finished. This water will remain in depressions, in footmarks, and behind small obstructions. For this water to remain on the surface the whole profile must be saturated to the surface, and this means that the free water zone also reaches the surface.

In these situations, drains placed at the above-mentioned spacings, and on profiles of 10- to 12-in. (250- to 300-mm) depth do have a useful function, and this is the removal of ponded surface water. The removal of this ponded water will occur at the rate of the hydraulic conductivity of the topsoil.

Springs or Wet Spots

Springs or wet spots sometimes occur in a facility. If it can be established that this is being caused by water moving down a slope in an underground aquifer, and if it intersects with the surface at the point in question, the strategic placement of a subsoil intercept drain will solve the problem.

Such problems are common in some types of terrain, but care should be taken to establish that this is actually the problem before drains are installed. In many cases wet areas on sportsfields in particular are caused by local depressions and appear possibly to be coming from an underground source, when in actual fact they are just a collecting point for surface water. If the wet area is caused by surface water, subsoil drains will be of very little help in solving the problem.

Once it has been ascertained that the problem is being caused by a spring, a cutoff drain should be installed. This drain must be dug, not only to intercept the flow of water, but it must be dug below the gravel bed that is carrying the water, as shown in Figure 8.1.

The pipe should be laid on a gravel or coarse sand as described in Chapter 9, and this material should extend above the level of the aquifer. The top of the trench can be covered with a heavier soil, provided it will not migrate into the filter material around the pipe. There is no need to have the filter material come right up to the soil surface.

Water collected by this type of cutoff drain should be piped off to a suitable disposal point. This may be an existing stormwater pipe, a swale, or an area of the facility where this water will not cause any damage. If the flow of water in the aquifer is high, then consult a professional hydrologist to establish maximum flow rates, and design the pipe size accordingly.

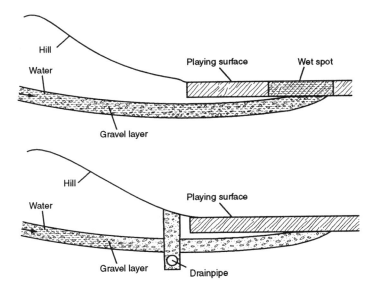

Figure 8.1. Water flowing through a subterranean aquifer and being intercepted and removed before it reaches the playing surface of the facility.

Designing a Subsoil Drainage System

Once the problem has been assessed and the decision has been made to install subsoil drains, the next step is to properly design the system. This design should follow a logical set of steps, as outlined in the following paragraphs.

1. Decide where to discharge the water collected in the subsoil drains.

If there is an existing stormwater system, open drain, creek, or pond in the area, decide if there is sufficient fall for the drains to discharge into it. A fall of 1 in 100 is usually the minimum that should be considered in such works.

Be aware of what happens when the drain, creek, etc. floods during high flow times, and make sure the discharge point of your system is above that level. If it is not, the system will not drain during wet weather, and there is a chance that water from the other system will flow up your pipes and make the problem much worse. This could occur at a critical time when the area needs to be draining at its most efficient rate.

If there is not an existing drainage system available to receive water from the facility, an area such as a swale or an area of waste land should be chosen to discharge the water onto. The most important thing to remember here is to not cause another problem by creating an unwanted bog from this discharge.

2. **Design the drain from the discharge point upward to the highest point to be drained.**

    This simply means that all the drains installed must fall at a gradient steeper than 1 in 100 toward the low point.

3. **Have a uniform fall on all main pipes.**

    All main lines should have the same grade on them between sumps, i.e., never have a change in grade without installing a sump, as a change in grade means a change in velocity of the water inside the pipe. Small particles of silt and sand are carried along in the pipe by the water. If the grade changes, say from 1 in 70 to 1 in 100 within the pipe without a sump being installed at the change of grade, as the water loses velocity it will drop some of its load of particles. These will deposit near the change of grade. If the pipes are small, e.g., 2.5 in. (65 mm), and are not precisely laid, there may be the potential for partial clogging of pipes.

    If there are laterals leading into a main pipe, the lateral may have a steeper slope than the main, but the lateral itself should have an even gradient over its whole length.

Once the discharge of the water has been taken care of, the next objective is to decide where to place the drains, i.e., the design of the system.

There are many situations where subsoil drains may be appropriate, but equally, there are a number of traditional uses of subsoil drains that have proved to be quite useless. Let us first look at the locations where subsoil drains work well and should be used.

Herringbone Pattern for Subsoil Drains

One of the most favoured designs for subsoil drains has been the *herringbone* pattern. This design usually involves a main line being laid with laterals coming off either side at regular intervals. The laterals join the main drain at an acute angle, usually between 30° and 45°. This angle ensures that there is not too much turbulence at the junction of the two pipes. If the angle approaches 90° the water entering the main pipe will cause greater turbulence at this point and this will reduce the ability of the main pipe to carry water at the rate it was designed to do.

This system allows there to be smaller-diameter laterals than the main pipe, which may need to increase in size as more laterals feed into it. (See Figure 8.2.)

The laterals are usually either flexible corrugated or solid smooth slotted pipes that allow water to enter it over its entire length. The main lines do not have to be slotted, but for smaller jobs the slotted pipe is usually used. For larger jobs where large volumes of water are being carried, solid pipes may be used. Cost is usually the main determining factor.

Note that the laterals do not join the main pipe opposite each other, as this also causes undue turbulence in the main pipe and reduces flow rates.

Determining the Size of Pipes Required

Pipe manufacturers supply performance charts for their pipes. These give maximum flow rates for a particular size of pipe at a given

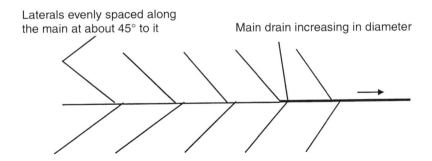

Figure 8.2. The herringbone design, with laterals feeding into a main drain, which has increased in size with the increased volume of water from all the laterals.

slope and for a particular length. As the length of the pipe increases, so too does the friction loss, and this decreases flow. Conversely, as the slope on the pipe increases, the flow rate increases. All of these factors must be taken into account when designing pipe sizes, both for laterals and for the main drain.

Most systems are designed so the laterals will run across the slope, and the main drain usually runs with the slope. As a general rule, the minimum slope on the laterals should always be 1:100. This ensures that water will always flow through the pipe even though there may be slight undulations in the pipe itself.

If the subsoil drains are being installed into a uniform slope of 1:70, and the main drain is installed parallel to the surface, the laterals will have a surface slope of only 1:100 if they are installed at 45° to the main, at a uniform depth.

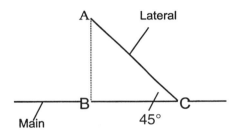

In the above example the ground has a uniform slope from left to right, and the fall on the main drain follows this slope and is therefore 1:70. Because there is a uniform slope in the direction of the main drain, then A and B will be at the same level. If the length of BC is 35 ft (10.67 m), then C will be 35/70 = 0.5 ft (0.15 m) lower than B.

This also means that A is 0.5 ft higher than C. However, the distance AC is

$$\sqrt{35^2 + 35^2} = 49.5 \text{ ft (15.1 m)}$$

A fall of 0.5 ft (0.15 m) in 49.5 ft (15.1 m) is 1:99, which is near enough to 1:100 for all practical purposes.

If the slope on the land were 1:100, the main line would have to be laid with a slope of 1:70, so that the laterals would be able to have a slope of 1:100 if they were laid at 45° to the main. This means that the main drain has to become increasingly deeper toward the outlet, to provide sufficient fall.

One point worth noting here is that the smooth slotted pipes can be laid at shallower grades than the slotted corrugated "agricultural" pipe. This is because it comes in rigid lengths and it does not get undulations in the pipes if they are carefully laid.

Always make sure that the main drain can deliver all the water it collects into another stormwater system through a sump, swale, dam, or creek. It is no good designing a system to collect water if there is nowhere for this water to go in high flow periods. These usually occur during prolonged wet periods, when other drains, creeks, and dams may have more water in them than normal.

Not all designs for subsoil drains have to be herringbone, but it is an efficient design. Single drains can be installed, or sometimes pipes can be laid in circles using the flexible corrugated pipe. Each situation should be judged on its particular set of circumstances, and the principles of design set out above adhered to.

The most important point to observe is that all laterals should be the same distance apart. This ensures uniform drainage of the whole area.

How to Size Pipes

Once it has been determined that there will be subsoil drains, and that they will be spaced at an appropriate spacing based on the calculations outlined in Chapter 7, the next step is to determine how large the pipes should be. This needs to be done for the laterals first, then the collector drain.

The best means of explaining the process is to work through an example (Figure 8.3). Let us assume that there is an area 147.6 ft (45 m) by 295.3 ft (90 m), with a uniform surface slope of 1:70 down the 295.3 ft, and no cross fall. Subsoil drains are to be installed at 6.6-ft (2-m) spacings. The topsoil depth is 12 in. (300 mm), and the estimated hydraulic conductivity of the soil is 2 in. (50 mm)/hr at 16 drops. (This is an actual design used at the Sydney International Equestrian Centre for the 2000 Olympics.)

There needs to be a rainfall event to design to, and in this case 2 in. (50 mm)/hr was chosen.

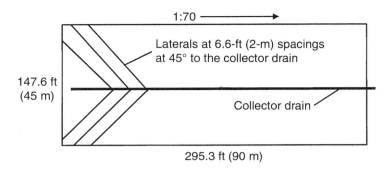

Figure 8.3. An area 147.6 ft × 295.3 ft (45 m × 90 m) with a central collector drain with laterals spaced 6.6 ft (2 m) apart at 45° to the collector pipe.

Sizing the Laterals

If the laterals are laid at 45° to the main collector drain they will have a slope of 1:100, as described above. This means that all the trenches will be a uniform depth.

The total area is 147.6 ft × 295.3 ft = 43,586.3 ft² (45 × 90 = 4050 m²). With drains spaced at 6.6 ft (2 m) apart, each drain will have a catchment of 6.6 ft (2 m) × the length of the lateral in feet to give an area in square feet (or in metres to give square metres).

If we now turn to a particular manufacturer's specifications for corrugated drainpipes, the maximum flow rate for a 2.5-in. (65-mm) pipe at a 1:100 slope is 0.26 gal (1 L)/sec, and for 4-in. (100-mm) pipe it is 0.83 gal (3.2 L)/sec.

The amount of water that actually reaches the pipe during a rainfall event must be estimated before the length of run for a particular sized pipe can be determined. From Hooghoudt's formula the rate of entry of water at the midpoint between the drains can be determined, but this is lower than the rate of entry of water into the pipe directly above itself, which will be at the rate of the hydraulic conductivity of the soil.

In the above example, the rate of drainage at the midpoint is:

$$\frac{4 \times 2 \ (1 \times 1)}{6.56 \times 6.56} = 0.18 \text{ in./hr}$$

or

$$\frac{4\times 50\ (0.3\times 0.3)}{2\times 2} = 4.5\ \text{mm/hr}$$

The rate at which the water reaches the pipe will be determined by the width of the trench. If the trench is 8 in. (200 mm) wide (this represents 10% of the total area of the catchment), then water will reach the pipe at an average rate of 10% of the hydraulic conductivity of the soil. In this case, this is 2 in./10 = 0.2 in. (5 mm)/hr.

It is worth noting here that while water is flowing into the drain at this rate from directly above the trench there will be virtually no water moving sideways into the pipe from the rest of the area.

For this reason, the figure calculated from the width of the trench, its proportion of the catchment area, and the hydraulic conductivity of the soil can be used to determine the maximum rate at which water will reach the pipe. In this case it is 0.2 in. (5 mm)/hr.

Rain at 1 in. over 1 ft^2 produces 0.108 gal of water. If the drain is draining the whole surface at an average rate of 0.2 in./hr, this will produce 0.120 gal/ft^2/hr. Therefore, if the 2.5 in. pipe with a maximum flow rate of 0.257gal/sec (926.6 gal/hr) draining a 6.6-ft-wide strip can be run for a maximum length of:

$$\frac{926.6\ \text{gal/hr}}{6.6\ \text{ft (spacing)}\times 0.120\ \text{gal/hr}} = 1170\ \text{ft}$$

Rain at 1 mm over 1 m^2 produces 1 L of water. If the drain is draining the whole surface at an average rate of 5 mm/hr, this will produce 5 L/m^2/hr. Therefore the 65-mm pipe with a maximum flow of 0.99 L/sec (3564 L/hr), draining a 2-m-wide strip, can be run for a maximum length of 356 m before it reaches maximum flow rate:

$$\frac{3564\ \text{L/hr}}{2\ \text{m (spacing)}\times 5\ \text{L/hr}} = 356\ \text{m}$$

In this instance the maximum run of any lateral can only be

$$\sqrt{73.8^2 + 73.8^2} = 104.4\ \text{ft}\left[\sqrt{22.5^2 + 22.5^2} = 32\ \text{m}\right]$$

because of the chosen design in Figure 8.3. Therefore 2.5-in. (65-mm) pipes can be used for all the laterals, and in fact will only be at one-tenth of their capacity.

Sizing the Collector Drains

Still using the above example, the total area of 43,586 ft^2 draining at the rate of 0.2 in./hr will generate 43,586 × 0.2/12 = 726.4 ft^3/hr × 7.48 = 5,434 gal of water per hour, or 90.6 gal/min, or 1.51 gal/sec.

In metric, the total area of 4050 m^2 draining at the rate of 5 mm per hour will generate 4050 × 5 = 20,250 L of water per hour, or 5.625 L/sec.

As a general rule, smooth pipes are quoted on internal diameter, and corrugated pipes are quoted mostly on external diameter. Make sure you get the manufacturer's flow rate chart for the product you are designing for.

From Table 8.1 it can be seen that a 4-in. (100-mm) corrugated pipe on a slope of 1:70 has a maximum flow rate of 1.0 gal/sec (3.8 L/sec). This pipe would not be large enough to run for the whole length of the collector drain, as 1.51 gal/sec (5.625 L/sec) was required.

A 4-in. (100-mm) corrugated pipe would, however, be able to cope with this flow for half of the distance of the collector pipe. An area of 21,793 ft^2 × 0.2/12 = 363.2 ft^3 = 2716.8 gal/hr = 0.75 gal/sec, which is below the 1.0 gal/sec shown in Table 8.1. (2050 m^2 × 5/3600 = 2.85 L/sec is below the 3.8 L/sec shown in Table 8.1.)

On the other hand, a 4-in. (100-mm) smooth pipe would cope with the whole catchment, as it has a maximum flow rate of 2.27 gal/sec (8.6 L/sec), which is well above the 1.51 gal/sec (5.625 L/sec) required.

The solution for the collector pipe in this situation is either a smooth 4-in. (100-mm) pipe for the whole 295.3 ft (90 m), or 4-in. (100-mm)-diameter corrugated pipe for about half the distance, and 6-in. (150-mm)-diameter pipe for the remainder.

Using these principles, it should be possible to size pipes for any proposed subsoil drain layout.

Flower Bed and Shrub Bed Drainage

There is a great deal of merit in installing subsoil drains in shrub beds and flower beds, particularly if they are large. If the bed is more

Table 8.1. Maximum flow rates in gallons per second (litres per second) for four sizes of smooth and corrugated slotted drainage pipes, at two different slopes, 1:70 and 1:100. (This information is from one manufacturer; others will be slightly different.)

Pipe Size diameter, in. (mm)	Smooth Slotted Pipe		Corrugated Slotted Pipe	
	Slope 1:70 gal/sec (L/sec)	Slope 1:100 gal/sec (L/sec)	Slope 1:70 gal/sec (L/sec)	Slope 1:100 gal/sec (L/sec)
2.5 (65)	0.71 (2.7)	0.61 (2.3)	0.32 (1.2)	0.26 0.99
4 (100)	2.27 (8.6)	1.90 (7.2)	1.00 (3.8)	0.85 (3.2)
6 (150)	6.82 (25.8)	5.68 (21.5)	3.06 (11.6)	2.59 (9.8)
8 (200)	14.8 (56.0)	12.15 (46.0)	6.74 (25.5)	5.68 (21.5)

Note: The diameter of drainage pipes varies, as some manufacturers may call a product 4 in. (100 mm) diameter, but this is external diameter, and the internal diameter may be only 3.5 in. (90 mm). Care should be taken when selecting pipes to make sure that the flow rates on the chart are for the internal diameter of the actual pipe you are installing.

than about 6 ft (2 m) wide, and is of any length in excess of 6 ft, then the bed may benefit greatly from one or more subsoil drains being installed in its base to remove water when the bed becomes saturated.

When these large beds are being constructed, it is prudent to install subsoil drains to ensure that the bed does not become waterlogged, particularly if such plants as bulbs are being grown and have to sit in the soil over a wet winter.

If there are no subsoil drains installed, the topsoil in the bed will have to rely on the raised edge of the bed to remove water. Hopefully, the base will also remove water, but as demonstrated in Chapter 5 this can be quite slow, and during prolonged wet weather the bed can remain very wet for long periods.

These conditions are disastrous for bulbs and for other small plants, which will perform badly in such situations. Shrubs that are susceptible to wet conditions may die during periods of prolonged wet weather. These waterlogged conditions are also ideal for such root diseases as *Phytophthora cinnamomi,* which thrives on wet, warm conditions.

The approach to draining beds with raised edges is different from the approach for those that are flush with the surrounding soil.

a. Raised beds

If the topsoil is raised above the top of the subsoil the edges of the bed will act as drains, and water will weep out the raised edges. (See Figure 8.4.)

This method can be used to ensure that the roots of any plants are kept well drained. It is commonly used with vegetable gardens and works equally well in any horticultural situation. Edges of the bed act as drains and water weeps out.

Let us consider what the drainage rate of the centre of a raised bed with no subsoil drains would be if the bed were 20 ft (6 m) across, the hydraulic conductivity of the topsoil were 2.36 in. (60 mm)/hr, and the topsoil depth were 12 in. (300 mm).

The midpoint would drain as if there were two drains spaced 20 ft (6 m) apart, i.e., the longest distance that water has to travel to a drain is 10 ft (3 m).

$$D = \frac{4 \times 2.36 \times 1 \times 1}{20 \times 20} = 0.023 \text{ in./hr}$$

$$\frac{4 \times 60 \times 0.3 \times 0.3}{6 \times 6} = 0.6 \text{ mm/hr}$$

If the top 4 in. (100 mm) were to drain to field capacity, then the drainage rate would be:

$$D = \frac{4 \times 2.36 \times 0.67 \text{ in.} \times 0.67 \text{ in.}}{20 \times 20} = 0.011 \text{ in./hr}$$

$$D = \frac{4 \times 60 \times 0.2 \times 0.2}{6 \times 6} = 0.27 \text{ mm/hr}$$

The average drainage rate for the top 4 in. (100 mm) will be:

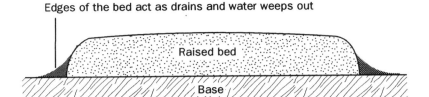

Figure 8.4. The edges of a raised garden bed remove water from the topsoil in the same way as a drain located at both edges of the bed. (See also Figure 4.6.)

$$\frac{0.023+0.011}{2}=0.017 \text{ in./hr}$$

$$\frac{0.6+0.27}{2}=0.435 \text{ mm/hr}$$

This sort of soil would have an air-filled porosity of about 30%, therefore the top 4 in. (100 mm) would contain 1.2 in. (30 mm) of water if the profile became saturated. It would therefore take about 70 hr to drain the top 4 in. (100 mm) of the profile. In a period of constant rain this is far too slow, and under these conditions plant roots would die and bulbs would rot.

If, however, there were one subsoil drain placed in the middle of the bed, the situation would dramatically improve. The two edges would still act as drains, and the drain would also work as an additional exit point.

One drain in the middle of the bed would decrease the "drain" spacing from 20 ft (6 m) to 10 ft (3 m), because the raised edges of the bed act as drains.

This has the effect of markedly increasing the drainage rate:

$$D=\frac{4\times2.36\times1\times1}{10\times10}=0.094 \text{ in./hr}$$

$$\frac{4\times60\times0.3\times0.3}{3\times3}=2.4 \text{ mm/hr}$$

The average drainage rate for the top 4 in. (100 mm) is 0.068 in. (1.7 mm)/hr. This is four times faster than the rate without the central drainage pipe. It means that the top 4 in. (100 mm) would drain in 17.6 hr. This is a very reasonable drainage rate, and would ensure that even after very heavy rain, or during long periods of rain, the top 4 in. (100 mm) of the topsoil in the bed with the one drain down the middle of it would always be near field capacity soon after the rain. This would ensure that the bed never became waterlogged, if you were relying on the drains alone.

If the bed were on a slope of, say 1:20, and were still 20 ft (6 m) wide, would one subsoil drain along the low side of the bed be sufficient? In this scenario (shown in Figure 8.6), any drainage back to the top edge is ignored because it would be such a small amount.

In this case we use only half of Hooghoudt's formula:

$$D = \frac{K \times H_1 \times H_2}{S^2}$$

Where H_1 = is the depth of the bed (1 ft) + the extra head, which is the depth due to slope, which is $1:20 \times 20$ ft = 1 ft. Therefore H_1 = 1 + 1 = 2. H_2 is the depth of the bed = 1 ft. The drain spacing S will be 20 ft because the water has to travel the full length of the bed to reach a drain. (See Figure 8.5.)

The top 4 in. (100 mm) will drain at:

$$\left(\frac{2.36 \times 2 \times 1}{20 \times 20} + \frac{2.36 \times 1.67 \times 0.67}{20 \times 20}\right) / 2 = \frac{0.0118 + 0.0066}{2} = 0.009 \text{ in./hr}$$

$$\left(\frac{60 \times 0.6 \times 0.3}{6 \times 6} + \frac{60 \times 0.5 \times 0.2}{6 \times 6}\right) / 2 = \frac{0.3 + 0.17}{2} = 0.235 \text{ mm/hr}$$

This is a very slow drainage rate. It would take over 5 days to drain the top 4 in. (100 mm) if the base did not drain, and is clearly far too slow. If the drain were placed down the middle of the bed it would drain at least four times faster, as was the case above.

All of these calculations depend on there being a low capillary fringe in the soil in the bed. If there is above 20% by weight of organic matter in the soil, and it is not compacted during construction, the above figures are valid. If, however, the soil is compacted during construction, the hydraulic conductivity will fall and the capillary fringe will rise. In this case the bed may never drain.

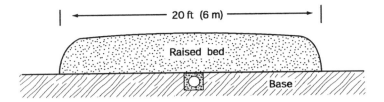

Figure 8.5. A raised bed with a drain in the middle of it. The two edges act the same as drains, so with this single drain added, the system behaves as if there were three drains present spaced at 10-ft (3-m) centres.

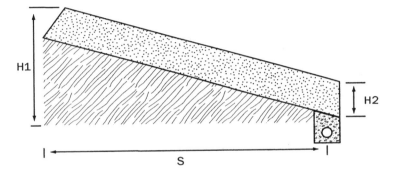

Figure 8.6. The difference between H_1 and H_2 when calculating the drainage rate in a bed with a slope on it versus the rate for a single subsoil drain at the bottom of the slope.

A key summary point here is that drainage is not just simply installing subsoil drains into beds. Care must be taken not to drive vehicles over them during construction. Any compaction caused by foot traffic should be relieved by raking and fluffing up the soil in the compacted areas. It is also essential to rip the base to ensure it drains as well.

b. Shrub and flower beds that are flush with the surrounding soil

When the bed is flush with the surrounding soil, where is the best place to locate the drains? Let us consider two options. (See Figure 8.7.)

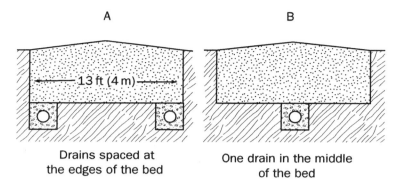

Drains spaced at   One drain in the middle
the edges of the bed   of the bed

Figure 8.7. A bed that is flush with the surrounding soil. In A, a drain is installed at either edge, and in B, one drain is installed in the middle.

In case A, if the topsoil had a hydraulic conductivity of 2.36 in./hr, and a depth of 1 ft, and the width of the bed was 13 ft, then the top 4 in. at the middle of the bed would drain at

$$D_{\text{Top 4 in.}} = \frac{\dfrac{4 \times 2.36 \times (1)^2}{(13)^2} + \dfrac{4 \times 2.36 \, (0.67)^2}{(13)^2}}{2} = 0.04 \text{ in./hr}$$

In metric for case A, if the topsoil had a hydraulic conductivity of 60 mm/hr and a depth of 300 mm, and the width of the bed was 4 m, then the top 100 mm at the middle of the bed would drain at:

$$D_{\text{Top 100 mm}} = \frac{\dfrac{4 \times 60 \times (0.3)^2}{(4)^2} + \dfrac{4 \times 60 \times (0.2)^2}{(4)^2}}{2} = 0.975 \text{ mm/hr}$$

This will be the same for case B, because the longest distance water has to travel to reach a drain is 6.5 ft (2 m). So in this situation, it would be cheaper and easier to run one line down the middle of the bed rather than have two lines down each side of the bed.

It is very important in the case of all flower and shrub beds for the base to drain, as this is the best insurance against waterlogging. A good-draining base will mean that the bed will drain more evenly, and the capillary fringe will also be drained. The base of these beds should be ripped, and gypsum added to the ripped surface at the rate of 10 lb/100 ft² (500 g/m²).

Racecourses

Most racecourses that are built today have conventional profiles, which usually have a topsoil depth of 12 in. (300 mm). Because there are such large areas involved in racecourses, it is common practice to only install subsoil drains in the first 33 ft (10 m) out from the rail. This is because there is usually a good crossfall on the track, and much water will be shed toward the rail. Close to the rail is also the area of intensive use, so with the combination of extra water and more use, drains in this area are good insurance.

The soil used should be a sandy loam, and should have a specification that meets all three of the following aspects of the specification:

1. Particle Size Distribution, or Mechanical Analysis

USDA Sieves (mm)	% Retained by weight
> 2.0	0
1.0–2.0	0–10
0.1–1	60–80
< 0.1	15–30
< 0.002 (clay)	2–5

2. Compacted Hydraulic Conductivity

At 16 drops, the hydraulic conductivity must exceed 2 in. (50 m)/hr, and at 32 drops it must exceed 0.6 in. (15 mm)/hr. As mentioned in an early chapter, this method has been developed by Dr. Jakobsen, and is described in Chapter 12.

3. Water Holding Capacity at 1 m Suction

The water holding capacity at 1 m suction must exceed 10 % by weight.

The combination of all three of these criteria will give a soil that will still drain after heavy use in the wet, will grow good grass, and will not be too droughty in the summertime.

The drains are usually spaced at 6.6-ft (2-m) centres, extending out from the rail for a distance of 33 ft (10 m). It is better to extend the drains further out from the rail if this is economically possible.

If the soil used has a hydraulic conductivity of 2 in. (50 mm)/hr, and the soil profile becomes saturated by prolonged rain, the above drain spacings would drain the top 4 in. (100 mm) as follows:

$$\frac{\left(\frac{4\times2\times(1)^2}{6.6\times6.6}\right)+\left(\frac{4\times2\,(0.67)^2}{6.6\times6.6}\right)}{2}=0.132 \text{ in./hr}$$

$$\frac{\left(\frac{4\times50\times(0.3)^2}{2\times2}\right)+\left(\frac{4\times50\,(0.2)^2}{2\times2}\right)}{2}=3.25 \text{ mm/hr}$$

These drains will drain the midpoint between the drains at 0.132 in. (3.25 mm)/hr. If the-air filled porosity of the soil is 25 %, there will be 1 in. (25 mm) of precipitation to be removed from the top 4 in. (100 mm). This will take 1/0.132 (25/3.25) = almost 8 hr.

This appears to be quite a long time to drain the top 4 in. (100 mm), but remember that the base will also be draining, hopefully at 0.04 in. (1 mm)/hr or better. It will then be draining at slightly better than 0.04 + 0.132 = 0.172 in. (4.25 mm)/hr, hence it will only take about 6 hr to drain.

This is an excellent drainage rate for such a soil. Once again it demonstrates that people should not put too much store in subsoil drains draining surfaces quickly. (See Figure 8.8.)

In this situation the base should also be ripped, and gypsum added to ensure that it drains at a satisfactory rate. Ripping the base ensures that the subsoil drains are only an insurance policy to cope with any areas where the base may only drain at a very slow rate. Regular decompaction of the base when it is dry with a verti-drain or vibrating subsoil decompaction machine also ensures that the base drains well and evenly.

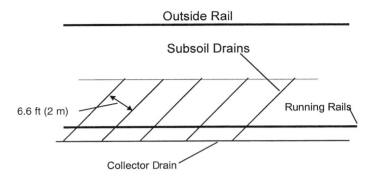

Figure 8.8. Subsoil drains installed on the inside 33 ft (10 m) of a race-course.

CHAPTER 9

Installing Subsoil Drains

Introduction

As discussed in previous chapters, many drains installed in sportsturf and garden situations simply do not work because of poor attention to design and poor understanding of soil drainage properties. Leaving aside this category of drain failure, a very high percentage of subsoil drains are installed incorrectly. As a consequence, after a short period of time they do not work, or only work very poorly. Usually this is because they are surrounded by the wrong filter material. (Engineers refer to the material that surrounds the pipe as filter material.)

Particle Migration into Coarser Gravels

The most common mistake is to surround the pipe with a coarse gravel or crushed stone similar to that used for surfacing roads. In some specifications, gravel as large as 0.4 in. (10 mm) is specified, but more commonly the "pea gravel" range is specified [usually perceived as being between 1/4 and 3/8 in. (or commonly in metric, 5 and 7 mm) in diameter]. Overall, these gravels are too large in diameter.

People have become so worried about the filter material entering the pipe that they have lost sight of the far more important issue of the adjacent soil particles migrating into the filter material surrounding the pipes.

If a coarse material is used to surround the pipe, the finer particles from the adjacent soil or sand will migrate into the voids of the

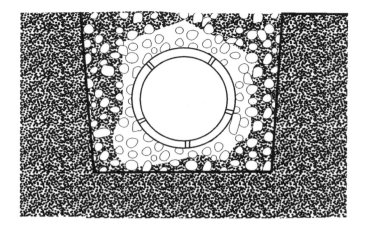

Figure 9.1. A slotted pipe surrounded with a coarse gravel, illustrating how the surrounding soil has migrated into the voids and reduced the hydraulic conductivity of the filter material to that of the surrounding soil.

material surrounding the pipe. As this occurs, the hydraulic conductivity of the filter material becomes the same as that of the adjacent soil. (See Figure 9.1.)

When pipes that were surrounded with a coarse gravel are dug up after a couple of years, it is often found that the voids in the gravel have been filled with the surrounding soil. Under these conditions significant amounts of fines can move into the pipe, with a great deal of this migration occurring with the first few waterings of the newly installed pipe. In other words, the damage is done very quickly, and the efficiency of the system can be quickly lost.

There is a tradition that has developed whereby a subsoil drain will be surrounded by a coarse gravel, say 1/2 in. (10 mm), and this material is then covered with a finer gravel or sand to prevent migration from the material above it. This practice is illogical, as it allows the soil particles to migrate from the sides and the bottom of the trench into the voids in this coarse gravel.

Often this causes the whole of the filter material surrounding the pipe to become contaminated with fines, the very thing the layer of sand over the coarse gravel is intended to prevent.

This coarse gravel surrounding the pipe serves no good purpose and adds nothing to the efficiency of the system compared to having a finer sand or gravel surrounding the pipe, which does not allow any particle migration from the surrounding soil.

Choosing the Correct Filter Material

The choice of the filter material to surround a drainage pipe should be made on the basis of the type of soil or sand that will cover and surround it. If the drain is in a shrub bed with a fine soil, the filter material surrounding the pipe should be a coarse washed river sand, and definitely not a gravel.

There is a very simple rationale in choosing this size of material. With a clean coarse washed river sand there will be virtually no particle movement into the filter material, and the sand will carry all the water that reaches it from the soil above at a rate 10 to 40 times faster than the rate water reaches it from the soil. The hydraulic conductivity of a good shrub bed soil may be 4 in. (100 mm)/hr, whereas the hydraulic conductivity of the sand material will be 120 to 160 in. (3000 to 4000 mm)/hr.

There is no logic in going for a gravel that may have a hydraulic conductivity of 800 in. (20,000 mm)/hr, as it achieves absolutely nothing more than the sand in conducting the water from the soil above to the pipe. However, if coarse gravel is used, it will allow the surrounding soil to migrate into the voids and severely decrease the hydraulic conductivity of the gravel itself.

Why risk making a system less efficient? Use a sand—not a gravel—to surround the pipe.

Table 9.1 is a specification for a clean drainage sand that can be used as a filter material to surround subsoil pipes in all soil situations.

It is very important to use a sand (or fine gravel) that has no fines, because fines can slow the hydraulic conductivity of the material significantly. The hydraulic conductivity of the drainage material should be in excess of 100 in. (2500 mm)/hr.

There is a situation where gravel is appropriate as the filter material to surround the pipes. This is when the material above the filter material is a USGA sand. Then the drainage pipe filter surround can be a gravel, but it should follow the specifications outlined in the *USGA Green Section Record,* 1993.

The most important feature of this is that the D_{15} of the gravel must not exceed the D_{85} of the sand root zone material by more than

Table 9.1. Recommended coarse washed sand to be used as a filter material around subsoil drains in soil. The specification is in both % passing Australian standard sieves, and % retained in USA Department of Agriculture sieves.

AS Sieve Sizes, mm	% Passing by Mass	USDA Sieve Sizes, mm	% Retained by Mass
9.5	100	> 8	0
4.75	98–100	4.0–8.0	0–2
2.36	70–100	2.8–4.0	0–20
1.18	30–78	2.0–2.8	0–20
0.600	2–15	0.5–2.0	55–100
0.300	0–4	0.10–0.5	0–10
0.150	0–1	< 0.10	0

a factor of 5 ($D_{15 \, (gravel)} \leq 5 \times D_{85 \, (root \, zone)}$—Bridging Factor). Also there is a Uniformity Factor ($D_{90 \, (gravel)}/D_{15 \, (gravel)} \leq 2.5$).

This gravel should be used for golf greens, bowling greens, and other perched water table sand-based structures, but should **not** be used for subsoil pipes in finer soils.

Table 9.2 shows what a specification for the gravel material discussed above could be.

There is a belief that if the gravel size is too small, it will somehow magically migrate inside the pipe and block it. These assumptions are wrong, and like a lot of other assumptions in drainage, are not based on fact and good science.

The facts are that the holes in most slotted drainage pipes are from 0.08 to 0.2 in. (2 to 5 mm) wide, and may be up to 0.5 in. (13 mm) long. The surface area of these slots is usually less than 1% of the total surface area of the pipe. These slots are also distributed around the sides of the pipe, so that probably only 30% of them would ever be oriented in such a manner that particles could fall into them, i.e., positioned in the top segment of the pipe.

The only way the gravel particles can enter the pipe is from above, and this is under the influence of gravity—particles would have to fall into the hole from above. Once the slots are positioned in the bottom two-thirds of the perimeter of the pipe, the particles cannot fall into them, as they do not move upward into holes.

Table 9.2. Recommended gravel to be used around pipes in a perched water table construction with a USGA sand.

Australian Standard Sieve Sizes, mm	% Passing by weight
9.5	100
6.7	90–100
4.75	80–100
2.36	0–25
1.18	0–5
0.6	0–3
0.3	0–2
0.15	0–2
0.075	0–2

So we now have a situation where there may be two to four lines of slots where the holes are oriented sufficiently toward the top of the pipe to allow some particles to move into the holes.

The popular argument from engineers is that the smaller particles in the gravel will pass down through these holes, and will continue to flow into the hole, causing cavitation and slumping above this hole, and of course blocking the pipe. But in fact this simply does not happen.

What actually happens when the gravel is dumped over the slotted pipe in the trench is that a few particles may drop into the pipe initially if the filter material is bone dry. However, after a couple have dropped in, the material above the hole bridges, and no more particles fall in. This bridging is caused by the irregular shape of the particles, the friction between the particles themselves, and the weight of the material above, all working to lock the smaller particles into voids in the gravel that is not near the holes. These fine particles can't magically find pathways through the network of different sized pores in the gravel simply to seek out the holes and then fall into the pipe. Once the particles have locked together and "bridged" above the holes, this prevents any further particle movement.

If the filter material has any moisture content at all—and this is almost always the case—not even the initial few grains will fall into the pipe. This is because the water molecules are held very tightly onto the particles and are also attracted to each other, with the result that this greatly increases the friction between the particles.

Under these conditions there is very strong bridging, and no particles will move. Once the filter material becomes wet, i.e., with the first wetting of the material, there will be no further movement, as all the particles completely lock together.

When talking about creating perched water table profiles with a uniform sand over a coarser gravel, the principle of preventing migration of the sand into the gravel shows how smaller particles do not migrate into the voids of the larger gravel underneath. Sands that meet the USGA specification have a D_{85} of between 0.4 and 0.7 mm, i.e., there will be 15 % of particles larger than this size. Using the USGA bridging factor, the D_{15} of the gravel—i.e., the smallest 15 % of the gravel—must not be smaller than 5 × (0.4 to 0.7 mm), or 2.0 to 3.5 mm.

When the sand and the gravel are matched in this way the sand will bridge and not migrate into the voids in the gravel below, even though many of these voids are larger than 0.4 to 0.7 mm, and up to 40 % of the sand particles may be smaller than these voids. In actual fact this factor of five is probably conservative, and there are some who believe that if the sand is moist there will be no migration with a factor closer to seven. There is no need to risk migration by going closer to ten, as the finer material will do the job perfectly well. The authors agree with the USGA factor of 5, and with perched water table constructions we always adhere very strictly to this standard. If the sand is bone dry it can migrate into the voids of even the recommended gravel, but if the gravel is coarser the risk of this happening is greatly increased.

The USGA has spent a considerable amount of money over the years doing research to come up with these recommendations, and it is naive for people to ignore these standards by saying "I think that coarser gravel works" without having these views backed up by hard data and good science. Please follow the standards set down by the USGA for these types of perched water table constructions for the selection of the gravel. If the gravel is too coarse, the bridging factor will be too high, and if the sand is very dry it will migrate into the gravel.

The same principle applies to selecting the correct coarse sand or gravel to surround pipes. Using the finer filter material prevents migration of particles from the surrounding soil from blocking up the filter material.

To demonstrate how little sand migrates into pipe slots, take a piece of slotted drainage pipe about 12 in. (300 mm) long and block off each end with plastic held with rubber bands so that no material can enter the pipe from the ends. Bury this in a clean coarse sand—as specified in Table 9.1—that has no fines, i.e., particles below 0.25 mm, and about 50% of the particles below 2 mm. Hence, more than half of the particles will actually fit through the slots.

The popular assumption is that the sand will simply pour into the slots on a continuing basis, with a pile of sand ending up under each hole. Nothing could be further from the truth.

If the sand is completely dry, i.e., oven dry, there will be a very small amount of sand entering the pipe. But the movement of sand stops quickly, and after an initial flow it stops completely. This experiment can be repeated with the pipe and the very dry sand being placed on a soil shaker and vigorously shaken for a minute. In our experiments, the amount of sand that entered the pipe was collected and weighed. About 80 g—or about 45 cm^3 of sand—had entered the piece of 4-in. (100-mm)-diameter slotted corrugated pipe, 12 in. (300 mm) long. This represents only 4% of the volume of the pipe, and thus could never block it. (See Figure 9.2.)

This was an extreme test that could never happen in the field, because the vibration was the main factor in moving the particles into the pipe. In reality, this caused sand grains to move into holes on the sides, as well as the top, and possibly some from the bottom. In a field situation such vibration would never occur. The sand was also oven dried, and this never happens in the field, as there is always some moisture in the sand.

The trial was repeated using sand that was only slightly moist, as is almost always the case in the field. Most of a stockpile of sand will be at field capacity, except for the sand within about 2 in. (50 mm) of the surface, which may be drier. The moisture content of the sand may only be 2%. If you place the back of your hand onto sand in this condition, a few grains of sand will adhere to it. This does not happen with oven dry sand. If the sand is quite wet, then a large number of sand grains will stick to the back of your hand.

When the sand had this very low water content, ***no particles migrated into the slots*** when the sand was placed over the pipe. When

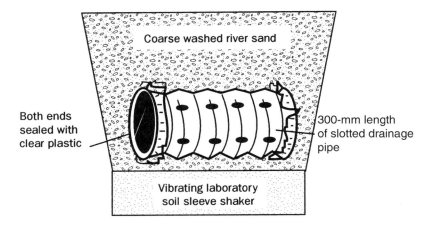

Figure 9.2. A piece of 4-in. (100-mm) slotted drainage pipe surrounded by a coarse washed sand, being shaken on a soil shaker. The trial was carried out with an oven dry sand, and also with the same sand slightly moist.

the pipe was subjected to the same vibration as in the earlier experiment, the number of particles that entered the pipe could be counted, and the volume was less than 2 cm^3. This represents less than 0.16 % of the volume of the pipe, which is hardly cause for concern.

What actually happens in the field situation is that the ever so slightly moist coarse sand forms a bridge across each slot, and no particles migrate into the holes in the drainpipe.

In summary, always remember: when installing subsoil drains in any situation other than under USGA sands, use a clean coarse washed sand, not pea gravel, and certainly never a gravel with a D_{15} that is more than 1 mm.

Installing Subsoil Drains

Subsoil drains should be cut into the area below the soil that is being drained. For example, if the top 12 in. (300 mm) of a topsoil is to be drained, the drain should be installed as shown in Figure 9.5. The points below should be followed carefully when installing subsoil drains:

1. Make sure the sides of the trench are clean and vertical. The bottom of the trench should be cleaned with a trenching

shovel to ensure that any heavy clay, or other material from the trench, is removed. If the trenching machine is leaving clay or subsoil material on the surface of the turf, a sheet of plastic should be laid on the surface to prevent contamination of the topsoil, as it will drastically reduce the beneficial properties of the topsoil. This material should be picked up and removed from the site, and not spread over the topsoil.

The bottom of the trench should have a uniform slope with no undulations. This can be checked with a dumpy level or simply by setting up above the trench a string line parallel to the theoretical line of the bottom of the trench. By using a stick with the depth of the trench below the string line clearly marked on it, the depth of the trench at any point can be accurately measured. (See Figure 9.3.)

It is not necessary to have the bottom of the trench absolutely free of small variations, but when the first layer of sand is placed in the bottom of the trench, the top of this sand layer on which the pipe is placed should be parallel to the string line, i.e., at the required slope, and of a uniform slope.

This simple method is illustrated in Figure 9.3.

2. Place the 2-in. (50-mm) layer of sand in the bottom of the trench. To ensure that the top of this sand layer is also parallel to the string line, simply make another mark on the stick 2 in. (50 mm) below the first mark, and test the depth, as shown in Figure 9.3. Smooth out any undulations to give an even gradient on which to lay the pipe.

3. Make sure the pipe has been unrolled and straightened before placing it in the trench. Because this pipe comes in coils, it will have a tendency to twist and rise if it is not properly straightened before it is laid. This is very important, because if this coiling is not completely removed from the pipe it can raise itself up, and this can cause the bottom of the pipe to have high and low points, which will restrict the flow of water.

4. Lay the uncoiled and straightened pipe in the centre of the trench. Lay the pipe with one row of slits on the **bottom** of the trench. This is important because the **water enters a pipe from the bottom,** not from the top of the pipe. (See Figure 9.4.)

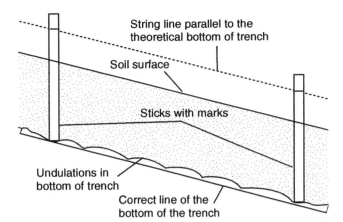

Figure 9.3. A simple method of determining if the bottom of the trench is at the correct gradient.

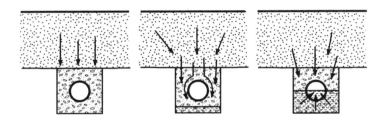

Figure 9.4. Water enters drainage pipes from the bottom via the surrounding filter material.

When the sand below the pipe becomes saturated, water will begin to enter it, and as more and more of the trench becomes saturated, water will begin to enter the slots higher up the pipe.

5. Cover the pipe with more of the same filter sand, making sure it completely surrounds the pipe. There should be at least 2 in. (50 mm) of sand covering the pipe in all parts of the trench. In most cases, this cover will become deeper as the pipe approaches its delivery point. (See Figure 9.5.)

6. Make sure the sand in the trench is firmed down. This can be done with the wheel of a tractor or by light rolling. It is

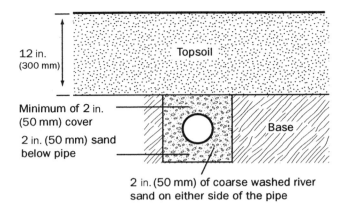

12 in.
(300 mm)

Topsoil

Minimum of 2 in.
(50 mm) cover

2 in. (50 mm) sand
below pipe

Base

2 in. (50 mm) of coarse washed river
sand on either side of the pipe

Figure 9.5. Subsoil drain cut into the base below the topsoil, with 2 in. (50 mm) of sand surrounding the pipe on all sides.

always preferable to have the sand extend slightly wider than the edges of the trench, and to have it slightly above the level of the base material. This reduces the possibility of the trench sand becoming contaminated with soil from the edges of the trench or by soil smeared over the surface of the sand.

Change of Grade When Laying Drainpipes

If there is to be a change in the grade on the bottom of the pipe that reduces the gradient, there should be a sump installed at this point. If water is flowing down a pipe and the gradient flattens out, the velocity of the water will be reduced. As water loses velocity, it reduces its capacity to carry a load. This means that if the water is carrying any sediment, i.e., very fine particles, some of these will be dropped out at the point where the gradient changes. This is why a sump with a silt trap should be installed—so when the material is dropped out of suspension it can be periodically removed from the trap where it collects.

It is not accepted practice to change the gradient of pipes in normal subsoil drains. However, in large herringbone designs, often the main pipe, which collects the water from the pipes that feed into it, may be at a slightly steeper gradient than the laterals. This has the added advantage of speeding up the delivery of water in the main line. It also ensures that any suspended material will be carried out of the system into a sump or an outlet.

Correct Use of Geo-Fabrics

Geo-fabrics should not be placed over the top of drains, as they are unnecessary if the correct sand has been used to surround the pipe. In horticulture and sportsturf the hydraulic conductivity of the soils is usually high, and under these circumstances there can be particle movement.

If the correct sand or gravel is used, these fine particles are washed through the filter medium to the bottom of the trench, and possibly into the pipe itself, where they are flushed away. This process of allowing the fines to pass through and be removed from the profile is how the system is designed. Geo-textiles will usually prevent these fines from moving through into the trench and the pipe.

A layer of fines can accumulate on the surface of the geo-fabric, and even though water may continue to move through it, it will drastically reduce the rate of flow. Geo-textiles are designed to prevent the movement of fine particles, particularly in low-flow situations, and they do this job very well. However, in sportsturf and horticulture in general they are unnecessary on top of the trench or wrapped around the pipe.

They are, however, very effective as a lining on the bottom and sides of a trench. This will prevent the movement of fines up into the drainage medium from a clay base. They also prevent the contamination of the drainage medium on the sides of the trench. So by all means use geo-textiles on the bottom and sides of subsoil trenches, as this will maintain the integrity of the drainage medium. But never put the cloth over the top of the pipe. (See Figure 9.6.)

Do not wrap pipes in geo-fabric, or place geo-fabric over the top of drains. Only use the material to line the sides and bottom of the trench or to cover the base.

Types of Pipes

There are two main types of slotted subsoil drainage pipes.

1. Flexible slotted, corrugated "agricultural" pipe

This comes in coils, and is simply rolled out and laid into the trench. As explained above, it is very important to ensure

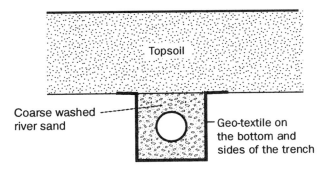

Figure 9.6. Subsoil drain with geo-textile covering the sides and bottom of the trench, but not covering the top of the sand in the trench.

that it is untwisted before it is placed in the trench. When laying the pipe always lay it with one set of slots on the bottom of the trench.

This pipe comes in sizes usually ranging from 2 to 8 in. (50 to 200 mm) diameter. The most commonly used size is 4 in. (100 mm), which with some manufacturers is the outside diameter, and the inside diameter is usually about 3.4 in. (86 mm).

Most pipes have from three to six rows of slots running longitudinally along their length. The size of the slots varies with manufacturers, but a common size is about 1/16 × 1/4 in. (1.5 × 6.35 mm). Some makes have the slots on the outside of the corrugations and running longitudinally, while others have them on the smaller diameter of the corrugations, and running around the pipe. These configurations all work and they all allow the water to freely enter the pipe.

The corrugations in the bottom of the pipe also have another function. There may be fines deposited in the bottom of these pipes in the corrugations, particularly soon after laying and after the first flush of water. As water flows along the bottom of the pipe, the corrugations cause turbulence, and this stirs up the fines. Some of this material will be washed over into the next corrugation, and so on. This fine material is progressively moved down the pipe to a sump. Hence these pipes will actually self-clean. It should be remembered that if the pipes are correctly installed with the correct filter material that suits the overlying topsoil, the amount of material to be moved within the pipe is very small. This process will not block pipes.

2. Smooth slotted pipes

These are rigid lengths of pipe with longitudinal rows of slots. They fit together with rebate joints where a flush exterior to the outside of the pipe is required, and spigot and socket joints where a deflection joint is required. Standard joints and fittings are available to allow these pipes to be joined at different angles.

These pipes have a higher flow rate than the flexible corrugated pipes, because they are smooth inside and there is less friction loss in the pipe. Corrugations create turbulence, which reduces the flow rate.

Standard flow charts are available from the manufacturers for all of these pipes, with flow rates dependent on the slope of the pipe, its diameter, and the length of run.

Rectangular Sectioned Pipes

There is a range of other drainage materials available, including those with a rectangular cross section. These pipes will always be less efficient than round pipes because there is greater friction loss, which is caused by a higher percentage of the pipe being in contact with the water inside the pipe.

If these pipes are used, check the manufacturer's flow rate specifications very carefully, because with some products the flow rate inside these pipes may be less than 40 % of that of the round section pipes.

Atlantis Draincell

There is an exciting new Australian product on the market that has been manufactured from recycled car batteries. This product has a radical new design. (See Figure 9.7.) This product can be used in place of conventional pipes in almost every situation.

The product is extremely easy to lay and in many cases does not require any trenching. It is excellent for drainage behind walls and roof gardens, and is now being used under golf greens.

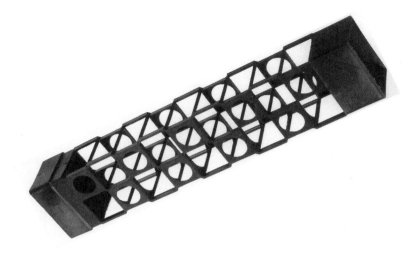

Figure 9.7. A section of Atlantis Draincell. These pieces lock together to form any length.

Outlet of the Pipe

As stressed already, the outlet of the pipe should preferably be into a sump, and the outlet into the sump should always be above the bottom of the sump. If the pipe is to discharge directly into a gully, or open drain, or just out into an open area, wherever possible there should be a small headwall built. This prevents the end of the pipe from being run over and crushed, as well as preventing surface erosion around and under the end sections of the piping. It also reduces the risk of the end of the pipe being covered over with soil or other material, which could eventually block the whole system.

Flushing Points

Some engineers insist on installing flushing points at the high points of subsoil drains. We believe that these are of no real benefit and are rarely if ever used. In any event, if the pipe becomes blocked and has so much foreign material in it that it needs flushing, then it has been incorrectly installed, and flushing will not help anyway.

If these pipes are installed using the correct sand around them, with an even grade along the bottom of the pipe, and if they have not been run over and damaged by heavy machinery, they should never become blocked. In summary, correct installation prevents drains from becoming blocked.

Joints

When subsoil drains are being joined, it is preferable to use commercial joints because they do not allow for one pipe to protrude into another. The common practice of cutting a hole in the pipe into which the feeder pipe is then pushed and wired or taped into place (or in some cases just pushed in) leaves a piece of pipe protruding inside the main pipe.

This causes turbulence inside the pipe in periods of high flow and can significantly reduce the flow rate of that pipe. If there are a number of feeder pipes protruding into a main, as in a herringbone situation, the capacity of the main to cope with its normal flow can be seriously reduced.

Most commercial slotted drainage pipe manufacturers have a range of fittings, which allow the pipes to be joined at a number of different angles with the pipes locking into the joints.

CHAPTER **10**

Surface Drainage

Introduction

Surface drainage is very important when designing, constructing, and maintaining a sporting surface or a horticultural facility, whether it is a golf course, racecourse, football field, park, school ground, flower bed, shrub bed, or even a lawn.

Some of the major problems of poor playability and performance of the above facilities are caused by poor surface drainage. These problems result from poor design, poor construction, or failure to repair damage and depressions with good maintenance.

To understand how and why water accumulates on the surface of the soil during rain or even irrigation, there is a need to understand the dynamic situations that occur as the water enters the soil and moves down through the profile. It would be appropriate here to go back and review the principles and processes of *infiltration, infiltration rate,* and *hydraulic conductivity,* dealt with in Chapter 2, and the *capillary fringe* from Chapter 5.

In summary, when water reaches the soil surface, either as rainfall or irrigation, it enters at a particular rate, which is called the *infiltration rate.* This rate varies, depending on two factors. The first is soil structure, as the rate will be slower in a heavy clay soil than a sandy soil. The second is the moisture status of the soil at the time rainfall or irrigation commences. If the soil is very dry the water enters quickly. If the soil is moist it enters more slowly. The rate of entry continues to slow down as saturation is approached. (See Figure 2.6.)

The initial rate at which water enters the soil is very fast because the water is filling the large pores near the surface as well as the smaller ones. As soon as these large pores are full, gravity begins to pull water downward. The rate at which this downward movement occurs will determine how much pore space is available for more rain or irrigation at the top of the soil profile. If the rate at which the rain is falling is faster than gravity can pull it further down the profile, the top of the soil will become saturated for a time. Refer to Table 2.1 to see how these rates vary for different soil types at different moisture status at the time of rainfall or irrigation. If water continues to reach the soil surface under these conditions, *runoff* or *ponding* on the surface will occur.

In the sportsturf and urban horticultural situation the profile is usually shallow. Topsoil ranging in depth from 3 in. (75 mm) to 8 in. (200 mm), and rarely 12 in. (300 mm), is placed over a clay subsoil base, which is often highly compacted. This shallow topsoil and clay base combination has a major problem, because once all of the topsoil becomes saturated, the rate at which the water can enter the soil is then determined by the rate at which the subsoil can remove it; i.e., very slowly, as this is the hydraulic conductivity of a much finer material than the topsoil. Hence, under these conditions runoff or ponding can occur quickly, often after only relatively small amounts of rain.

Saturation

It is important to understand the relationship between a saturated profile and surface runoff, hence the need for surface drainage. Refer to Figure 2.5, which shows that saturation is a dynamic process, always changing with time as water drains through the profile. From this figure it can be seen that the location of a saturated zone is continually changing in a soil profile during and after rain.

Water can pond on the surface, and runoff will occur when only the top of the profile is saturated. In this situation runoff will continue while the rate of rainfall exceeds the hydraulic conductivity of the topsoil.

Runoff will also occur when the saturated zone reaches a much slower-draining base if rain continues. The profile fills up completely, and again water ponds on the surface. Under these circumstances runoff will continue while rain falls at a rate higher than the hydraulic conductivity of the base, which in most circumstances is very slow. So when the profile is saturated to the surface, almost all the rain that falls thereafter must be removed by surface drainage or, if there is no surface slope, water will pond.

Ponded water on the surface that persists for more than a couple of hours indicates that the whole of the topsoil is saturated. This usually happens in depressions and on flat surfaces.

In both cases there is a need to remove this ponded water on the surface if the facility is to be used shortly after the rainfall event or to prevent waterlogging killing plant roots. The only way to remove this water is to have a slope on the soil surface, because if the surface is flat the water will remain there until it slowly drains down through the profile or is evaporated from the soil surface. If this rainfall event has occurred during the winter there will be very little evaporation and this ponded water can remain there for days, depending on how slowly the subsoil drains.

If a base drains at say, 0.02 in. (0.3 mm)/hr, it would take 83 hr to remove 1 in. (25 mm) of ponded water from the soil surface (without any evaporation). To ensure there will be no damage caused by usage it may take another couple of days to drain the top 2 in. (50 mm) of topsoil back to field capacity and mechanical stability.

Need for Surface Drainage

If all sporting and horticultural facilities were flat there would be no runoff. Every drop of rain that fell onto the surface would have to drain down through the topsoil and then through the much slower-draining subsoil to enable the facility to be used or to grow plants. It is for this reason that almost all successful football fields, golf fairways, racecourses, and flower and shrub beds have some surface slope on them to shed excess water during heavy or prolonged rain.

There are two components of surface drainage. The first is to have sufficient slope to make the water flow away from the facility at a reasonable rate. If the slope is too steep the rate of flow is too high, and erosion will occur. If the slope is too small, the water does not move quickly enough.

Secondly, when surface water is made to flow off a desired area it must be collected in some sort of drain and removed from the site, otherwise the problem of the ponded water has only been moved from one place to another.

Effect of Slope

What is the correct amount of slope on a surface to effectively remove water? This amount will vary depending on what the area is being used for, and in many cases what the end user will allow, particularly in the case of sportsfields. In the case of golf fairways, for example, some golf course architects like to have a minimum slope of about 1:60 on all their fairways. Such a slope on a soccer field would affect the roll of the ball and is not acceptable. The turns on many racecourses have quite steep slopes—up to about 1:16—to help the horses run around the bend at high speed without running off toward the outside rail, but much gentler slopes of 1:70 are preferred on the straight sections. These slopes also effectively shed surface water.

Poor Design

There are many sections of golf courses, racecourses, sportsfields, parks, school grounds, shrub beds, and lawns that are in poor condition because they have been wrongly designed with insufficient surface slope.

If the design is correct, the surface slope is adequate, and the facility is properly constructed, most of the above problems disappear. Less water remains on the surface, playability increases, and plant roots do not remain waterlogged for long periods. Less damage occurs after rain and maintenance costs fall.

Because almost all sporting facilities are built from soil that will readily compact, if water reaching the playing surface is not rapidly removed by surface slope, the topsoil will become saturated. Once in this state it becomes vulnerable to damage. The process that causes this damage is called ***puddling,*** which happens when play occurs on surfaces where the top of the profile is saturated. Under these conditions the fines become mobile and are redistributed within the soil by being forced into the voids. This process causes compaction, as it increases the soil density and reduces the pore space. As a consequence, the hydraulic conductivity of the soil is reduced, and there is less air space in the soil for good grass growth.

When the soil is compacted there are two things contributing to poor drainage. Firstly, the hydraulic conductivity is lowered, reducing both the rate at which water is removed down through the soil as well as reducing the infiltration rate, which makes this area reach runoff quicker than adjacent uncompacted areas. Secondly, the capillary fringe is increased in height (as explained in Chapter 5), which makes the top of the profile remain saturated for long periods of time.

As a result of puddling, the area becomes unacceptable for play, as it will be continually wet and muddy, particularly in the wintertime. This situation is a surface drainage problem. Subsoil drains rarely fix these problems. You need to get it right at the design stage, as remedial action is often limited and costly.

There are vast amounts of water that have to be removed from the surface of sports facilities during prolonged rain. For example, 0.5 in. of rain falling on 2 ac generates 27,154 gal (12.7 mm over 0.81 ha generates 102,870 L). There can be dire consequences if these amounts are not removed quickly and efficiently. Water remaining on the surface after the top of the soil profile is saturated must be removed as quickly as possible to ensure that the mechanical stability of the soil surface is achieved. If it remains on the surface it must be drained down through the profile, which is a very slow process. Rapid removal of surface water enables the soil surface to quickly become aerated and stabilised naturally, as the soil profile drains downward due to gravity.

Surface drainage has the advantage that large volumes of water can be discharged. As the precipitation rate increases, the depth of water on the surface increases and the rate of surface flow increases. An even surface slope without depressions is desirable, and the length of slope should not be too long. For sporting facilities, if the length of the slope is too long, water will accumulate after about 76 yd (70 m) and will cause problems.

The water that runs off due to surface slope should be removed by drains of some sort as quickly as possible. One strategy is to collect this water in dish drains with sumps at regular intervals, then pipe it to a stormwater system. This ensures a minimal depth of ponded water, a drained surface soon after rain ceases, and a rate of flow that never reaches a velocity that will cause erosion.

For sportsfields a surface slope of 1:70 is found to be acceptable for play and also sufficient to remove excess water quickly, provided the length of any slope does not exceed 76 yd (70 m) in any one direction. Golf fairways do not have such a slope restriction, but the minimum of 1:70 should prevail, although much steeper slopes can occur. These steeper slopes will help remove surface water more quickly from the playing surface.

The relationship between the surface slope S (mm); depth of ponded water D (mm); and the velocity of flow; V (m/sec) is given in the following equation:

$$V = 0.35 \times D^{0.67} \times S^{0.5} \text{ m/sec}$$

This equation is valid for a bare surface only. A dense grass cover would restrict the water flow considerably at high velocities, but at very low velocities the equation may give useful estimates of the depth of ponded water to be expected during rainfall of a known intensity.

Using, for example, a surface slope, S = 1:70 and a depth of water, D = 4 mm gives a water velocity of 0.1 m/sec. A 1-m-wide cross section would yield

$$0.1 \times 4 \times 60^2 = 1440 \text{ L/hr}$$

In the same example, if the depth of ponded water (D) were 0.16 in., the water velocity would be 4 in./sec, and a 3.3-ft cross section would yield 380 gal/hr.

Rain in the amount of 1 mm gives 1 L of water per m^2, so at the bottom of a 70-m-long slope the water depth would be 4 mm when the runoff is

$$1440/70 = 20.6 \text{ mm/hr}$$

Further up the slope the depth of water would be less than 4 mm and the velocity would be less than 0.1 m/sec.

A water depth of 10 mm gives a water velocity of 0.2 m/sec, giving a discharge of 50 to 100 mm/hr.

The above figures are in metric, but the reader may choose to convert them to inches and U.S. gallons to use a more familiar set of units. (1 mm = 0.039 in.; 1 L = 0.264 US gal.)

It can be seen from the above figures that there are great benefits from having a good surface slope, which quickly removes surface water. Huge amounts of water can be shed by properly designed surfaces. Insufficient slope will mean a large percentage of this water will have to pass down through the profile, often taking weeks for soils to dry out, particularly in the winter. If the soil cannot accept this water, it will remain on the surface. This is usually a completely unacceptable situation. On the other hand, facilities with similar soils that have adequate surface slope will be playable quite quickly, because only a fraction of the water actually enters the soil. As explained above, most damage is caused to sports surfaces when they are used in the rain, or when the surface is saturated, causing puddling and compaction and their associated problems.

These areas remain wet, particularly in the winter, and quickly become mud heaps. These muddy areas often develop in depressions where surface water remains after rain. They often appear toward the end of a long slope, usually in excess of 76 yd (70 m). High-traffic areas at the ends of long slopes are prime targets for this type of damage. (See Figure 10.1.)

The problems shown in Figure 10.1 occurring on football fields and golf fairways are caused by high traffic on areas at ends of long slopes. Large volumes of water are reaching these areas. Once the surface is broken they become very uneven, and even more water collects there.

These areas tend to quickly get larger, particularly when they begin in high-use areas of a football field, a high-traffic area on a golf course, or near the rails on a racecourse.

An example of this occurred in Canberra in 1978 when an Australian Rules football ground was badly damaged (90 % of the surface became a mudheap) because three matches were played in continuous heavy rain over a seven-hour period.

On a local golf course, an area at the end of a cart path became a quagmire from foot and cart traffic after one weekend's use when wet. There was insufficient slope on this area to shed surface water, so it collected there. On other similar areas on the course where there was sufficient surface slope and the surface water drained away quickly, little damage occurred even though there was high traffic.

On a local racecourse an area in the straight was always of concern as it was often wet, causing the cancellation of several meetings. There was a theory that there was a spring under that area of the track, but a detailed survey of the area showed that there was a de-

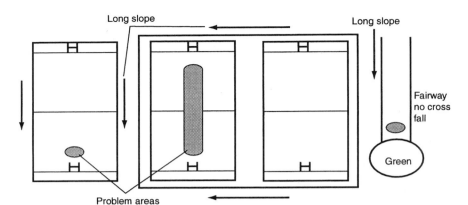

Figure 10.1. Damage-prone areas on football fields and a golf fairway where high use occurs at the end of a long slope.

pression and that the surface slope on the area was much less than the adjacent areas. Once this was corrected with topdressing, making the surface slope similar to the adjacent areas, the problem disappeared because surface water was shed at the same rate as the adjacent areas.

The removal of surface water quickly reduces the possibility of puddling and the subsequent damage that occurs.

There are two aspects of surface drainage, and these are:

1. The rate at which the water is running over the surface during rain.

To illustrate the effect of slope and length of slope on the discharge of surface water, we will assume a situation where the topsoil is near saturation and a further 2 in. (50 mm) of rain then falls over 4 hr. Over 2.47 ac (1 ha) this rain generates 130,000 gal (500,000 L) of water which has to be discharged from the surface. This is an enormous volume of water. To put it in context, it is the amount of water that has to be removed from only one average golf fairway or a football field during this heavy rainfall event.

Calculations of the depth of water flowing over the surface at different degrees of slope and length of slope (using the above equation) are shown in Table 10.1.

Table 10.1. Depth of water in inches and millimetres, flowing over surfaces with different slopes during rain falling at 0.5 in. (12.5 mm)/hr.

	Length of Slope		
Slope	**76 yd (70 m)** **in. (mm)**	**110 yd (100 m)** **in. (mm)**	**164 yd (150 m)** **in. (mm)**
1:70	0.114 (2.9)	0.142 (3.6)	0.177 (4.5)
1:100	0.125 (3.2)	0.157 (4.0)	0.20 (5.1)
1:150	0.142 (3.6)	0.177 (4.5)	0.224 (5.7)

These calculations assume a bare soil surface, but a grass cover would have only a small effect on water flow at these slow rates, thus giving us a good guide.

The velocity of flow over the 1:70 slope is 16.4 ft (5 m)/min, while the depth of water on the surface is 0.114 in. (2.9 mm). It will therefore take 14 min for water to travel from the top to the bottom of a slope 76 yd (70 m) long. After the rain has stopped, the velocity of the flow slows as the depth of water above the soil surface lessens. The field should be fairly clear of surface water after half an hour to one hour.

The velocity of flow over a 1:150 slope is 13.1 ft (4 m)/min, with a depth of water on the surface being 0.142 in. (3.6 mm). It will take 18 min for water to travel 76 yd (70 m). Because of the deeper water and the slower flow rate it will take at least 2 hr for the surface water to clear in this case.

This difference does not appear to be very significant, but when considered in the context that almost all playing surfaces do not have uniform slopes, it becomes very important. It should be noted from Table 10.1 that the length of the slope contributes more to the depth of surface water than the degree of slope. On a slope 70 m long, there is only an increase in the depth of the water of 0.7 mm as the slope changes from 1:70 to 1:150. However, on a slope 150 m long the depth of water increases by 1.2 mm as the slope changes from 1:70 to 1:150.

This is very important to understand when designing facilities, because if there are long slopes—greater than 76 yd (70 m)—the depth of water that builds up at the ends of these slopes during and after heavy rain is unacceptable.

2. The effect of the amount of water left on the surface after rain ceases.

In the above example it was assumed that the surface was perfectly even and without depressions. However, it should be realised that sportsturf areas are not perfect because they have small undulations and depressions. It is very important to realise how much water is retained in these surface depressions after the surface flow has ceased. Water also accumulates in these areas before runoff starts.

For uneven surfaces this stagnant surface water (water that is not flowing or moving) is more important than the depth of running water, because it is around for a long time, compared to the very short periods when water is actually running over the surface.

This ponded or stagnant surface water only drains downward through the profile at the drainage rate of the base. This rate may only be 0.4 in. (1 mm)/hr or less when a heavy clay subsoil base is restricting the rate of drainage. This means that 1 in. (25 mm) of surface water will take 25 hr or more to drain away. In many cases where the base drains more slowly this could take several days to drain.

Let us consider the effect of a 0.2-in. (5-mm)-high "wall" on areas with different slopes (Figure 10.2). On a 1:70 slope such a wall would retain water for a distance of 14 in. (0.35 m) behind it, while for a 1:150 slope water would be retained out to a distance of 30 in. (0.75 m). With the 1:150 slope, the surface area of the stagnant water is more than twice as large as with a 1:70 slope. The volume of water left on the surface to slowly drain down through the profile is also doubled. A wall of 0.2 in. (5 mm) is a very small obstruction on the surface of a playing field or golf fairway.

Walls are created by uneven topdressing, wheel ruts made by service vehicles, footprints made by players, indentations made by golf carts, and sometimes just by bad construction methods.

When water lies on the surface for long periods it creates a situation where the risk of further walls being formed is greatly increased. A surface with no walls exceeding 0.2 in. (5 mm) may be considered a nearly perfect one, as stagnant water 0.4 to 0.6 in. (10 to 15 mm) deep may be found even on very good sportsfields after prolonged rain.

The steepness of slope is most crucial when it comes to the removal of stagnant water, and a slope of 1:70 is considered necessary

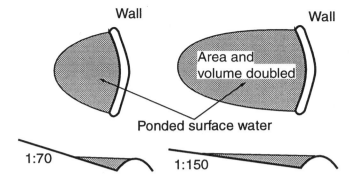

Figure 10.2. The difference in size and volume of ponded water behind a small 0.2-in. (5-mm) "wall" for slopes of 1:70 and 1:150.

to minimise the amount of water that remains behind walls after surface flow has ceased.

The length of slope is also very important because, as explained above, with long slopes; i.e., in excess of 76 yd (70 m), the bottoms of these slopes are wet for longer periods than areas further up the slope. The largest amount of running surface water reaches the bottom of the slope and the flow continues longer. Soft areas develop here because they are always the last to dry; consequently more damage tends to occur in these areas.

The amount of water that collects in a depression may greatly exceed the amount of rain or irrigation water that falls on this area, because the depression is being filled up with water from neighbouring areas. This is why depressions become soft and remain wet long after adjacent areas have drained and become firm again.

After the surface runoff has ceased, deep drainage through the profile is necessary to remove any stagnant surface water as well as excess water in the topsoil, particularly during winter or periods of low evaporation. It is essential for this to occur to provide sufficient air space in the root zone of the grass. Air is essential for the survival and development of roots.

The amount of stagnant water can be kept to a minimum on all surfaces, including lawns, and shrub and flower beds, as well as sporting surfaces, by having the correct surface slope together with a very even surface. Commonly, however, as much as 0.4 to 1 in. (10 to

25 mm) of water is ponded on many sportsfields in small depressions and behind walls; these cause the problems outlined above.

Once the surface water has been removed, the top of the topsoil is still saturated. It is necessary to drain at least the top 2 in. (50 mm) to ensure surface stability, which greatly reduces damage caused by play. The drainage of the top of the profile of shrub and flower beds is also very important because if the plant roots are waterlogged for too long, this can lead to unthrifty plants or even death.

What is an acceptable surface slope?

What is an acceptable surface slope for the various sports? A slope of 1:70 is an acceptable surface slope for top-grade sport, and certainly for all lower-grade users. It is worth noting that some major sports stadiums in Australia have a surface slope of 1:70. These include Suncorp Stadium in Brisbane, where international soccer and rugby league are played. On this ground there is no slope longitudinally up the centre of the ground, and the fall is to the sides of the field. This slope has not affected the users and the players seem unaware of it. The Sydney Cricket Ground, which is used for international cricket and Australian Rules football, had surface slopes of 1:70 for many years and was always considered a very good-draining field because surface water was shed quickly. A recent reshaping of this surface reduced the surface slopes to less than 1:100, and the result has been disastrous, with severe damage occurring with football use during the winter. As long as the original slope existed there were no complaints from the footballers.

In the late 1970s a survey of some 250 sportsgrounds was carried out in Canberra to determine which performed the best during wet winters. Those that performed the best after prolonged rain all had good uniform slopes of about 1:70 and less than 76 yd (70 m) in any one direction. These were the most efficient at shedding surface water. Based on this information all new sportsgrounds constructed in Canberra have had surface slopes of 1:70 and no slope more than 76 yd (70 m). The performance of these facilities has produced a dramatic fall in the amount of money required each spring for renovation. These facilities have had less play cancelled due to wet conditions than most of the other older fields, which have flatter slopes and/or very long slopes in the one direction.

The old engineering rule of thumb of 1:100 for surface slopes for sportsfields is definitely not enough. 1:70 is ideal for water removal with a turfgrass cover and still allows for first-class sporting activity. Slopes of greater than 1:70 tend to begin to influence play—players begin to feel they are running up- or downhill, and the ball tends to roll too much downhill in sports such as soccer.

A slope of 1:70 is probably the minimum slope that should be used for golf course fairways, but in many cases the slope is much steeper. It is important in golf fairway design or redesign to make sure that water is shed off the fairway fairly quickly. This should be toward the outside of a fairway. Long fairways should never slope for more than 60 to 70 yd (55 to 70 m), because wet areas develop after this distance.

Prevent water from reaching the playing surface or horticultural feature. When designing a sports surface, a lawn, or even a shrub bed, make sure there is no water flowing onto it from surrounding areas. Very often, otherwise well-designed facilities have problems caused by water flowing onto them from banks, parking lots, or surrounding areas, such as adjacent fairways on golf courses. (See Figure 10.3.)

Usually this water is easily diverted from the playing surface, lawn, or garden by installing surface cutoff drains to collect the water before it reaches the vulnerable area. Small diversion banks can also be used to deflect water. These are cheap, effective, and can be made part of the landscape. Subsoil drains are completely useless in these situations.

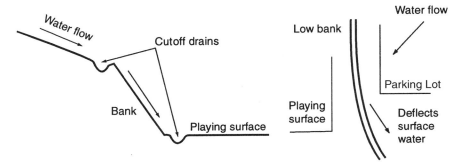

Figure 10.3. Water being prevented from reaching playing surfaces by using surface drains or banks.

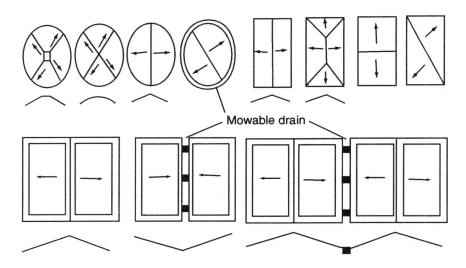

Figure 10.4. Some possible shapes for bases and location of mowable drains.

Design for Playing Fields

To design playing fields using the 1:70 slope and no slope longer than 76 yd (70 m) criteria, there are many possibilities that can be used, depending on the sport to be played and the slope of the site to begin with. Engineers should keep these in mind when doing the primary shaping of the base. It is usually just as cheap to provide a 1:70 slope as a 1:100, particularly when large machines are being used. The configurations in Figure 10.4 may be used.

It is also very important on golf courses to make sure that water does not flow all the way down long fairways. They should be tilted or domed to deflect water off the playing surfaces. If this surface water is not removed, areas will become boggy and damaged in wet weather. The same applies to greens and particularly to greens surrounds. (See Figure 10.5.)

Usually these problems can be solved by changing surface slopes, *not* by subsoil drains.

Mowable Drains and Swales

Mowable drains or swales are cheap, easily maintained, and effective. They are essentially surface drains that collect water and trans-

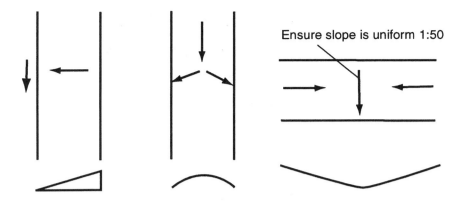

Figure 10.5. Slopes on fairways to deflect surface water.

fer it to underground pipes—usually through grated sumps—or to dams and creeks. The grass cover in the drain is usually the same as the surrounding playing surface and is mown with the same mowing equipment. The bottom of mowable drains must be easily mown by the mower that mows the sportsfield or golf fairway or surrounds.

Typical locations for mowable drains for sportsfields are shown in Figure 10.4 around the perimeter of ovals and in between fields. For golf courses they can be between fairways, around greens, or anywhere there is a need to collect excess surface water quickly and dispose of it into some other part of the stormwater collection system, such as dams, creeks, or into pipes via sumps. (See Figure 10.6.)

The bottom of the drain *must* be very uniform in slope, and should have a minimum slope of 1:50. This is essential to allow the very rapid removal of surface water. If there are even small depressions in the bottom of the drain, water will lie there, machines will bog, grass will grow long, rubbish will accumulate, and the drain will become a mess. The length of the slopes running into the grated sumps should not be more than about 22 yd (20 m), otherwise the tops of the sumps will end up too far below the surrounding area. This means sumps can be spaced about 44 yd (40 m) apart if the slope goes toward them in both directions.

A small concrete, oblique V-shaped strip, about 12 to 16 in. (300 to 400 mm) wide, in the middle of a mowable drain works extremely well and ensures an even slope on the bottom of the drain and rapid delivery of water to sumps. This, of course, does not suit all locations, but

Mowable swale with water flowing
into a grated sump

Figure 10.6. Longitudinal and cross sections through a mowable drain.

where it can be used it works much better than just a grassed bottom of
the drain because there are no small depressions to collect water. It is
also very good at being able to shape the sides to a definite edge, making construction with a uniform grade on the drain very easy.

The water collected in these mowable drains and swales should
be disposed of through solid pipes from the sumps to an existing
stormwater outlet.

Solving Surface Water Problems

Surface drainage is probably the most important aspect of
sportsturf, golf course, and general horticultural facility design. It is
the cause of more turf management problems than any other factor,
as it affects all surfaces and soil profiles if it is not adequately ad-
dressed.

Remedial action for many surface drainage problems is often quite
straightforward. Before rushing out and installing subsoil drains for
those wet area problems, look for a surface water problem. If you
have a wet area on a golf course or sportsfield, carry out the following
analysis of the problem to see if it is being caused by surface water.

1 If it is a sportsground, get a grid survey done with levels
   taken every 20 to 30 ft (7 to 10 m). From this data a
   contour map of the area can be generated, showing all the

surface falls and depressions. This process can identify a series of problems.

2. Check that there is no water flowing down a bank that surrounds part of the facility or from further up the hill.
3. Make sure there is no water being diverted onto the playing surface from a drain outlet or diversion bank.
4. Get out there when there is heavy rain falling and water is flowing over the surface. This is a very important thing for every manager to do at some time for the whole of the facility. You will often be surprised to find that water is flowing onto an area from some adjacent area. This may be another fairway, a parking lot, or a bank.
5. If there is surface water lying on an area during or immediately after heavy rain, take some pegs and a marker pen and peg the outside of the depression. Place one peg in the middle of the water or at the deepest part and mark it at the top of the water level. After the water has drained away you have two pieces of very valuable information. You know how deep the depression is and you also know precisely how big the depression is.

After this information has been collected, establishing that there is a problem being caused by unwanted surface water, the following remedial action can be taken.

Depressions

Once it has been established that there is a depression, the best way to fix it is to progressively topdress it out. This is best done during the growing season of the grass being used on the facility. Remember, never apply topdressing to more than 40 % of the height of the grass in any one application. Allow the grass to grow through the first application, then apply another if required. With deep depressions it may be necessary to topdress in several applications. This may need to be done over a couple of years to completely fill the depression.

If the depression is deep and is on an important part of the surface it may be better to lift the existing turf, add soil to fill in the depression, and then replace the turf over the new soil. It may be necessary to use new sod if the old grass was badly damaged. This area will need light topdressing to attain good levels once the sod has been established.

Depressions develop in such high-use areas as soccer goal mouths. As the season continues they become deeper, collect water, and become mud heaps. A simple strategy here is to slightly mound these areas at renovation. These areas should be ripped, have extra soil added to restore levels, then be rotary hoed, lightly compacted, and shaped so there is a fall away from the high-wear area. The slope on these areas can be 1:70 to quickly shed surface water. After returfing or oversowing they may need some light topdressing to ensure even slopes. This strategy has proved very successful in dramatically decreasing these bad depressions in Canberra. Players are much happier with a slightly mounded grassed area rather than a muddy hole in front of the goal. The same strategy works on golf tees.

Water Flowing Down Banks

If there is water reaching the playing surface from a bank, there are two possible remedial approaches. If there is a problem with disposing of water from the bottom of the bank, a cutoff surface dish drain should be constructed at the top of the bank to collect water before it reaches the bank. This means that any water that falls on the bank itself will still flow onto the playing surface. If the bank is a cut, in a cut-and-fill construction of the facility, which is common, this will probably suffice.

If the drain is placed at the bottom of the bank, this is the most desirable solution. The water collected by such a surface cutoff (dish) drain must be removed either by the slope on the bottom of the drain or by installing sumps that deliver the water to an underground stormwater pipe. Such sumps should be at regular intervals and preferably be grated, and flush with the ground to ensure user safety.

Water Flowing Across the Surface from Adjacent Areas

This is the most common of all of the surface drainage problems. We see it on football and baseball grounds where water flows from surrounding land, from under seating, off parking lots onto the playing surface, and even from other fields in the complex. It occurs on golf courses where water flows from one fairway to another, or from adjacent land onto the course.

There are two solutions. First, the water can be collected in cut-off dish drains, mowable drains, or swales as explained above. Secondly, small diversion banks or mounds can be constructed to divert the flow of water.

The diversion mounds are very simple and inexpensive to construct. First determine exactly where the water is coming from, and more importantly, where you can divert it to. If there are areas that do not get use and this surface water can be diverted onto them without causing problems, simply build the bank so the water is directed to those areas.

If there are no areas where the water can simply be relocated, then more complex solutions are necessary. The water may be diverted by the bank to a low spot where a sump can be constructed and the water collected can be piped away to a stormwater system or into a dam or creek.

On golf courses the use of low mounded diversion banks can be very effective in preventing water flowing across one fairway to the next. These banks do not look unsightly and have no effect on play. They are also cheap to build, as they can be shaped with poor soil. Provided they are topped with topsoil for grass growth, they are easily managed. The shape of these banks should be compatible with appropriate mowing equipment.

In many cases small diversion banks are the cheapest and easiest solution to the problem of diverting water from where it is not wanted. *Remember that water on the surface is a destroyer, and good surface slope is a money saver.*

CHAPTER **11**

Slit Drainage

What is slit drainage?

As outlined in Chapter 10, when sporting facilities, such as football fields and golf course fairways, are constructed using heavier soil that compacts, considerable damage can occur when play or use occurs when the surface soil is saturated. This damage is usually caused by the failure to remove surface water. One remedial method of quickly removing surface water is by the installation of slit drains.

This practice of slit drainage has become very popular in Australia over the past few years as a means of "fixing" poorly drained sportsfields, golf course fairways, and some racecourses that use warm-season grasses. The method entails a series of narrow vertical trenches being dug into the field and backfilled with sand to the surface. Some have pipes in the bottom of the trench, others do not. Usually these trenches connect with larger collector trenches, which have pipes in the bottom to carry the water away to an appropriate stormwater outlet.

Some of these systems work for a while and then fail. Others work poorly from the start. The best will work well for many years. It is probably worth pointing out here that none of them will work forever. This means that if you have a facility sand-slitted or trenched, it will need further work on it after a period of time—often every couple of years—to ensure that it continues to work.

Sand-slitting or trenching is a method of *surface drainage.* It is not a subsoil drainage system, even though there are pipes below the soil. The principle of this method relies almost completely on surface slope to move water sideways on the surface of the soil toward the trenches, which have a permeable sand brought right up to the surface. When the surface water reaches the top of the drain it moves quickly down through the sand to the pipes, which deliver it off to a stormwater system.

The drains themselves also act as subsoil drains; however, the amount of subsoil water they remove in this role is minute compared to the volume of surface water they remove.

When installing slit drains, the closer they are spaced the more efficiently they work. Conversely, if they are spaced too far apart, the soil in between the trenches will become saturated and puddle, causing muddy areas. This mud then becomes smeared over the top of the trenches, rapidly slowing down the entry of the water into the drains. On a normal football field this will occur if slits are spaced farther apart than 6.6 ft (2 m). If trenches are spaced at 3.3 ft (1 m) apart there is a very good chance that most of the surface water causing the damage will be removed quickly. On golf courses the drains can be spaced farther apart, but care should be taken to ensure that the pipes in the bottom of the slit trenches are large enough to be able to accept the water from the increased catchment area.

Design and Planning

When planning to install slit drainage into a facility there are a number of important steps that should be taken.

1. First find out where the surface water that is causing the problem is coming from. If there is water flowing onto the ground from an adjacent area, collect this type of surface runoff before it reaches the playing surface. You should only be dealing with water that falls on the playing surface itself.
2. Determine the rainfall event you wish to design the drainage system to cope with, e.g., this may be to remove 2 in. (50 mm) of water in one hour. This decision is vital for the future success of the operation.
3. Make sure there is a suitable stormwater system that the drainage water can be directed into and that there is sufficient fall for the collection pipes to get the water into it.
4. Design the surface trenches to run across the slope so as to maximise the efficiency of the delivery of the water into the slits.

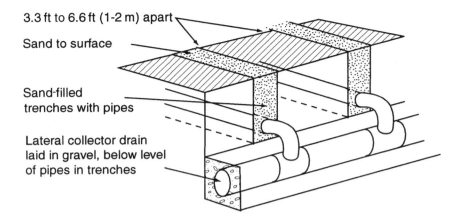

3.3 ft to 6.6 ft (1-2 m) apart

Sand to surface

Sand-filled
trenches with pipes

Lateral collector drain
laid in gravel, below level
of pipes in trenches

Figure 11.1. A typical cross section of a sand trench showing how it links with a larger collector drain.

5. These surface trenches should never be more than 6.6 ft (2 m) apart on high-use areas such as football fields or racecourses or high-traffic areas on golf courses. The distance the surface water has to travel to a slit must not be too far, as even small undulations and footmarks will inhibit the flow.
6. Once you have decided on the slit drain spacings, calculate the amount of water each of these slits will collect at the rainfall event you have chosen.

The basic design of a slit drainage system should also include a drainage pipe in the bottom of *every* trench. If we consider a trench that is 3 in. (75 mm) wide and 20 in. (500 mm) deep, when this is backfilled with a sand that has a porosity of 40 % it will only hold 3 in. × 20 in. × 12 in. × 40 % = 288 in.^{3}/231 = 1.25 gal per linear foot of trench. Every linear foot of trench can only hold 1.25 gal of water. If the slit drains are spaced 3.3 ft apart they will be filled by only 0.6 in. of rain, or if the slits are spaced 6.7 ft apart they will be filled with only 0.3 in. of rain.

7.5 cm × 50 cm × 100 cm × 40 % = 15000 cm^3 or 15 L per metre of trench. Every linear metre of trench will only hold 15 L of water. Slit drains spaced 1 m apart will be filled by only 15 mm of rain, or if they are spaced 2 m apart, they will be filled with only 7.5 mm of rain.

What happens if there are no pipes in the bottom of the sand slit trench?

There are those who advocate installing these trenches without any pipes in the bottom of the primary trenches. These sand-filled trenches then intersect with feeder sand-filled trenches having slotted pipes in them. The collector drains may be spaced as wide apart as 50 ft (15 m).

Let us consider this scenario where there are **no pipes** in the bottom of the feeder trenches, which are generously spaced at only 3.3 ft (1 m) apart, feeding into collector drains with pipes spaced at 50 ft (15 m) apart. Let us assume that the trench is 20 in. (0.5 m) deep and is filled with a sand with a hydraulic conductivity of 40 in. (1,000 mm)/hr (a coarse USGA sand).

If rain is falling at the rate of 0.6 in. (15 mm)/hr or more, the trench will fill with water and become saturated to the surface. Once this occurs, the maximum rate that water can ever flow in this trench when it is saturated to the surface (using Hooghoudht's formula) will be:

$$\frac{4 \times 40 \times 1.67 \times 1.67}{50 \times 50} = 0.178 \text{ in./hr}$$

$$\frac{4 \times 1,000 \times 0.5 \times 0.5}{15 \times 15} = 4.4 \text{ mm/hr}$$

This means that once the trench is full it can only cope with further rainfall at 0.178 in. (4.4 mm)/hr, which is a low rate of rainfall. So even with the slits spaced at 3.3 ft (1 m) apart, once the trenches become full, this system can only remove water from the surface at the low rate of 0.178 in. (4.4 mm)/hr. Any rain that is falling at a rate higher than that will simply run off. These systems are supposed to be designed to prevent runoff and to prevent water building up on the surface. Yet, the area in between the drains will become saturated. When runoff occurs, the areas in between the drains will absorb much more water. If they are played on while they are saturated, they will become damaged.

So you can see from this that just putting slit trenches into the ground without pipes in the bottom of the trench—even at 3.3-ft (1-m) intervals—is not the answer. It will initially collect the water quickly at the rate it is falling, but the trenches will fill up very quickly, even after as little as probably 0.8 in. (20 mm) of rain, and then take a very long time to move this water to the collector drain.

Even placing interceptor drains with pipes in them at intervals of 33 ft (10 m) will not work much better because the drainage rate in the trench without the pipe will only be 0.4 in. (10 mm)/hr. For rainfall events of more than about 1 in. (25 mm), runoff will be reached. Where there is prolonged rain the trenches will become saturated quickly. If the ground is used in this condition, damage will occur.

The biggest problem with this scenario is that when the heavier soil becomes saturated it becomes unstable and moves quite easily. This movement causes smearing, i.e., moving silt and clay particles over the top of the sand trenches, immediately and drastically reducing the hydraulic conductivity of the sand in the top of the trench. As this smearing continues it seals the top of the trench, reducing the rate of entry of water to that of the surrounding topsoil, making the slit drains useless.

This process also allows silt and other fine particles to be washed down into the sand in the trench, thus contaminating it. At best this will slow down the rate of movement of the water through the sand. At worst it will prevent any water getting into the sand trench.

There is another problem with particle movement down through a trench that has no pipes in its bottom. The fines that enter the sand at the top of the trench migrate to the bottom and form a layer. This layer could seriously impede any downward drainage through the bottom of the trench into the subsoil.

The drainage of water downward into the subsoil through the bottom of trenches without pipes cannot be dismissed, as under many circumstances more water may be removed this way than is drained sideways into lateral pipes. When trenches are cut into a playing surface, usually between 14 and 18 in. (350 and 500 mm) deep, they very often break through compacted bases that were formed when the facility was constructed using heavy machinery. This may mean that the base of the trench will drain at maybe 0.04 to 0.08 in. (1 to 2 mm)/hr. From the calculations above it can be seen that water being removed from the bottom of the drain at the rate of 0.08 in. (2 mm)/hr will be making a very large contribution to water removal, given that lateral movement of water in the sand may be slow, often less than 0.12 to 0.16 in. (3 to 4 mm)/hr. If the sand in the trenches has too many fines, its hydraulic conductivity may only be 12 in. (300 mm)/hr, or if the slits are spaced wider at 6.6 ft (2 m) apart, the base may be removing as much water as the linking lateral pipes. This, of course, will be very slow.

Obviously, anything that occurs to impede the rate of drainage of the base—such as the buildup of fines washed down to the bottom of the trench—will further slow an already slow system of drainage.

The authors consider that drains in the bottom of all slit trenches are essential to remove the amounts of surface water we experience from rainfall events in eastern Australia. You may be able to get away with no pipes in slit trenches in climates where rainfall is only light; but where it is of high intensity and may rain for periods of time at rates above 10 to 15 mm/hr, the slits have to be close together and the collector drains probably less than 33 ft (10 m) apart. Very careful planning must be done to ensure that such a system will work, even in climates where the rainfall intensity is low.

What happens if there are pipes in the bottom of the sand slit trench?

There is an enormous effect on the efficiency of the collection and delivery of water entering the slits when there are pipes in the bottom of the slit drain trenches. In Australia some contractors have machines that cut trenches 3 in. (75 mm) wide, and install 2-in. (50-mm) slotted pipes into these trenches. These pipes have a delivery rate of 7.9 gal (30 L)/min at a slope of 1:100 in the bottom of the trench. They are backfilled with a sand that usually has a hydraulic conductivity of 40 in. (1,000 mm)/hr or more. These slits are often spaced 6.6 ft (2 m) apart, and sometimes—if the client can afford it—at 3.3 ft (1 m) apart.

If the design is for Sydney the rainfall event may be 2 in. (50 mm)/hr, or for Melbourne, where the rainfall events are less, it may be 1.2 in. (30 mm)/hr. This rate must be decided upon as an essential part of the design.

Let us take the scenario of the higher-rainfall event, 2 in. (50 mm)/hr, and the wider slit spacings of 6.6 ft (2 m).

At 2 in./hr every 1 ft of trench will have to accept 6.6 ft × 1 ft × 2/ 12 ft or 1.1 ft^3 or 8.46 gal/hr (or 0.14 gal/min). The 2 in. pipe has a maximum discharge rate on a slope of 1:100 of 7.9 gal/min.

At 50 mm/hr every 1 m of trench will have to accept 2 m × 1 m × 50 mm or 200 cm × 100 cm × 5 cm = 100,000 cm^3 or 100 L/hr (1.67 L/min). The 50-mm pipe has a maximum discharge rate on a slope of 1:100 of 30 L/min.

How far apart should the main collector pipes be?

The next question in a design is how far apart should the collector pipes be, or conversely, how long should the run of a 2-in. (50-mm) pipe be before it feeds into a collector drain?

Each foot of pipe is accepting water at a rate of 0.14 gal/min and the maximum discharge rate is 7.9 gal/min; the maximum distance the pipe can run is then 7.9/0.14 = 57 ft. This means that there need to be collector drains spaced every 57 ft to collect the water.

Each metre of pipe is accepting water at a rate of 1.67 L/min and the maximum discharge rate is 30 L/min; the maximum distance the pipe can run is then 30/1.67 = 18 m. This means that there need to be collector drains spaced every 18 m to collect the water.

These calculations depend on the following:

- the slope on the bottom of the trench
- the size of the pipe in the bottom of the trench
- the catchment of the trench (spacing)

If the slit trenches were wide enough to have 4-in. (100 mm) pipes in the bottom of the trench, then in the same scenario as above—i.e., the slit drains spaced at 6.6 ft (2 m) apart—the length of runs would be increased. The delivery rate of a 4-in. corrugated pipe is about 50 gal/min, so 50/0.14 = 357 ft. Collector drains would only have to be spaced 357 ft apart.

The delivery rate of a 100-mm corrugated pipe is about 190 L/min, so 190/1.67 = 114 m. Collector drains would only have to be spaced 114 m apart.

It can be seen that this type of system—with drainpipes in the bottom of the trenches—will remove surface water ***continuously*** at the rate of 2 in. (50 mm)/hr (or whatever the design criteria were) for however long the rain continues to fall at or below that rate.

On the other hand, a system installed without pipes in the trenches is very limited in the amount of water it can remove after the initial rainfall event fills the trench. ***Pipes in all trenches are essential.***

It is false economy to install systems without pipes. For those people who tell you that pipes are not needed, use the information provided above and do your own calculations, and prove them wrong for yourself. Two-inch (50-mm) pipes are cheap, and as we have shown you, quite long runs can be installed with fewer collector drains. When you cost out the installation of drainage trenches without pipes keying into collector drains with 4-in. (100-mm) pipes, spaced at less than 33 ft (10 m) apart, there is not much difference in cost for a system with collector drains spaced at twice the distance apart with 2-in. (50-mm) pipes in all the slit trenches.

How to Calculate the Size of the Collector Drains

We have now determined the spacing of the collector drains. The next task is to determine the size of pipe. (See Figure 11.2.) In the above scenario the collector drains are spaced at 57 ft (18 m) apart, the slits are 6.6 ft (2 m) apart, and the rainfall event was 2 in. (50 mm)/hr.

As calculated above, the slit drain collects 8.46 gal of water per hour per linear foot. So every hour 57 ft collects 482 gal of water, or 8 gal every minute. If the length of the area being slit drained is 76 yd (228 ft), there would be 35 slit drains (one every 6.56 ft). 35×8 gal/min = 280 gal/min. From Table 8.1, it can be seen that either an 8-in. corrugated pipe or a 6-in. smooth slotted pipe would suffice, as both have a maximum discharge rate of 5.68 gal/sec or 341 gal/min, which will cope with the 280 gal/min.

There would be no need to use this size of collector pipe for the whole 76 yd, as a 4-in. pipe would suffice for some of the length. A 4-in. corrugated pipe can accept 51 gal/min, and as each 57 ft length of slit trench is generating 8 gal/min, the first seven slit trenches can feed into a 4-in. pipe and thereafter into a 6-in. pipe, which will accept a further thirteen. The remaining seventeen would have to feed into an 8-in. pipe.

If smooth slotted pipe were used it would need 4-in. pipe for the first fourteen, and 6-in. pipe for the remainder.

In metric, the slit drain collects 100 L/hr/m. Every hour 18 m collects 1800 L or 30 L/min. If the length of the area being slit drained was 70 m there would be 35 slit drains (one every 2 m). 35×30 L/min = 1,050 L/min. A 200-mm corrugated or a 150-mm smooth slot-

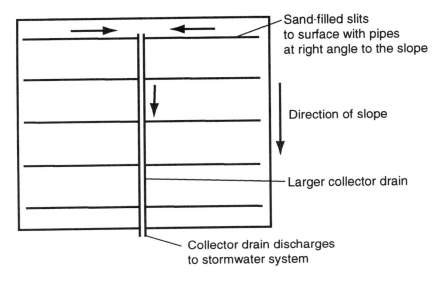

Figure 11.2. A typical design with laterals running across the slope feeding into a larger collector drain running down the slope.

ted pipe will suffice, as they have a maximum discharge rate of 46 L/sec or 1,290 L/min.

A 100-mm collector could be used for the first seven trenches, and thereafter into a 150-mm pipe for the next thirteen. The remaining seventeen would have to feed into 200-mm pipe. If smooth slotted pipe were used it would need 100-mm pipe for the first fourteen, and 150-mm pipe for the remainder.

The huge difference is that the system with pipes in the slit trenches is very efficient; the other is very slow and inefficient and will smear over and contaminate much quicker.

Use of Geo-Fabric in the Trenches Below the Pipes

We would recommend that a strip of geo-fabric be placed in the **bottom** of the trench prior to the pipes being laid if the subsoil is unstable. This is cheap and prevents silt and fine material moving up into the pipes and possibly blocking them; it is a very good and cheap insurance policy. The geo-fabric should **never** be wrapped around the whole pipe or placed on top of the pipe.

Type of Sand to Be Used in the Trenches

The sand used in the slit trenches has to be carefully selected and tested to conform to rigid engineering standards. Any old sand will not do. If the wrong sand is used in these trenches it will definitely lead to trouble, particularly after the first year. If the sand is too fine it will not have a high enough infiltration rate and its hydraulic conductivity will be too low. If it is too coarse the surface will not have enough traction and it will be unstable. It will be very difficult to establish grass on, and will be very droughty in the summer. The edges of the trench will break if the sand is too unstable. This contaminates the sand.

If the sand has a wide particle size distribution it will compact. This may greatly reduce the hydraulic conductivity.

If the sand is the correct size it will have a high infiltration rate and a high hydraulic conductivity. It will not compact and grass will grow reasonably well on it. The sand should meet the following specifications by weight:

mm	%
> 2	0
1–2	0–10
0.5–1	5–30
0.25–0.5	40–90
< 0.25	5–25

USGA sands fit this specification.

Perched Water Table in the Trenches

There is often a problem in establishing grass on the trenches, particularly if Bermuda (couch) or kikuyu is not being used, as the sand in the top of the trench tends to dry out and becomes quite droughty in the summer. This can be overcome by placing a 3- to 5-mm gravel layer in the bottom of the trench and covering it with about 8 to 10 in. (200 to 250 mm) of sand, depending on how fine the sand is. (See Chapter 9, Using the USGA Bridging Factor.) See (Figure 11.3.)

The gravel layer is often difficult to install, and many contractors will tell you that it is not necessary. If the trench top is not stabilised quickly with grass, this sandy trench can become a hazard, particu-

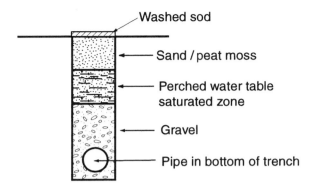

Figure 11.3. A trench with gravel in the bottom that perches a water table and prevents the top from drying out excessively in the summer time.

larly for footballers, who can easily turn an ankle. If the grass does not quickly take, the edges of the trench become vulnerable and break inward, and the fine material from the surrounding soil contaminates the sand, reducing the hydraulic conductivity.

Grassing the Top of the Trenches

After the slit trenches have been laid and backfilled with sand, or gravel and sand, there is always a problem in establishing a grass over the whole trench. If the area is grassed with Bermuda (couch) or kikuyu and the work is carried out during the summer, there is probably little that needs to be done, as these grasses will fill in over the trench quickly. By the next winter there will be a good grass cover to prevent the trenches from becoming hazardous.

If the trenches are to be covered by sod (turf), *only washed turf must be used.* If there is any soil on the turf the infiltration rate of the top of the trench will be dramatically reduced and the trench will be almost useless. Washed sod will "take" quickly and well on the sands used in these types of trenches; it doesn't matter whether the grasses are cool-season or warm-season, providing the warm-season grass is laid during its growing season and with sufficient time to knit before it becomes dormant.

Sprigging the trenches with stolons of Bermuda (couch) or kikuyu works very well in both coastal and cool-season grass areas, providing it is done during summer.

Sowing seed is difficult in these types of sands and requires constant watering. Fertilising needs to be carried out every two weeks until the grass is well established.

The addition of peat moss at the rate of 1 % by weight in the top 4 in. (100 mm) of the sand is often very beneficial in assisting the establishment of sod or stolons, and it is essential for the establishment of a grass cover when seed is used.

Removal of Soil from Adjacent to the Trenches

When the trenching machine digs the trench it is imperative that all the material dug from the trench be removed from the site and that the edges of the trenches are sharp and clean. Great care should be taken to ensure that none of this soil from the trench is spread back over the sand once the trenches are backfilled, as this completely defeats the purpose of installing this type of drain.

When the trenches are backfilled the sand should slightly overfill the trench. This ensures that none of the edge of the trench can collapse back into it, even slightly, as this will contaminate the sand and dramatically slow the drainage. (See Figure 11.4.)

Once the trenches have been consolidated by rolling, and by watering heavily, any excess sand above the desired level should be gently and carefully spread over the surrounding turf.

One strategy that works very well is to remove the existing sod (turf) from the line of the trench using a sodcutter that is, say, 12 in.

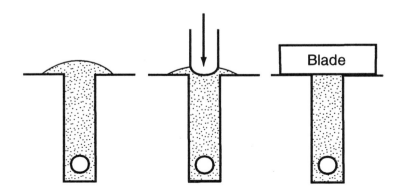

Figure 11.4. Sand slightly overfilling the trench, then compaction and levelling.

(300 mm) wide. The trenching machine digs the trench, say, 3 in. (75 mm) wide. Once the pipes have been installed and the trench has been filled with sand, lay the new 12-in. (300-mm) washed sod back into the area removed by the sodcutter. If there is a slight depression because of the sand removal, topdress the newly laid sod with the sand used in the trench to restore levels.

This process has the advantage of providing a neat finish and prevents the edges of the trench from being damaged and contaminating the trench.

Summary of What to Do, and What Not to Do, When Considering Having Slit Drainage Installed

1. Do not have trenches farther than 6.6 ft (2 m) apart.
2. Insist on having drainage pipes in *all* trenches. Cutting costs by only backfilling trenches with sand won't work— it only wastes money.
3. Have a proper design done by someone with experience in these designs (using pipes), taking into consideration the rainfall events with which you wish to cope.
4. Make sure the trenches run across the slope.
5. Only use an approved sand that has been tested by a reputable laboratory.
6. If geo-fabric is used it should only be on the bottom of the trench, not covering or surrounding the pipe.
7. Never cover the trench with sod (turf) having any soil on it. If sod is used it must have all the soil washed off it.

Remember that slit drainage is a form of surface drainage whereby the surface water has to flow down into a sand-filled trench. It is imperative to understand that there must be no impediments to the flow of water through the grass covering this sand, or the system will not work efficiently.

Future topdressing of this area must be done with the sand that was used in the trenches, because if a finer soil is used the drains will block up and all the money spent installing the system will be wasted.

After a number of years it will be necessary to take the top off the sand slits. To do this remove the sod (turf) with a turf cutter and discard it, as it will be contaminated with fines. Remove probably the top 2 in. (50 mm), replace the sand with new clean sand with the same specification as the original sand, and reestablish the grass cover as described above.

Remember: slit drainage is a form of surface drainage.

Where can slit drainage be used?

Slit drainage is a remedial process that can be used on existing facilities that have surface drainage problems. This process has been used successfully on football fields, other sportsfields, golf courses, and racecourses.

Sportsfields

Slit drainage can be very successful in areas where there are long slopes in excess of 76 yd (70 m) and areas where damage is occurring because of failure to remove surface water.

One point must be made here, and that is there needs to be surface slope for this process to work. A slope of 1:100 is probably the minimum needed, as the surface water needs to flow toward the slits at a reasonable rate.

Slit drains work well where there are large expanses of fields that have the surface slope in the one direction.

In some situations where there are insufficient funds to have a full slit drainage program to prevent water flowing from one field onto another, and the construction of a swale in between the two areas is not an option, possibly because of player safety, there is another solution.

A larger slit drain can be constructed in between the two areas, as shown in Figure 11.5.

Let us assume there are two areas, each of 2.47 ac (1 ha), side by side with a uniform slope running right across the two areas. A 15- to 18-in. (400- to 450-mm)-wide trench is dug, which is 18 in. (450 mm) deep at the shallow end, and there is a slope of 1:100 on the bottom of the trench to remove the water. The profile should be as follows: 2 in. (50 mm) of a 3- to 6-mm gravel, as described in Table 9.2; an 8-in. (200-mm) smooth slotted pipe, surrounded by and covered with a minimum of 2 in. (50 mm) of gravel; a 10-in. (250-mm) layer of a good USGA sand with 1 % peat moss in the top 4 in. (100 mm); and *washed sod* placed over the top of the trench.

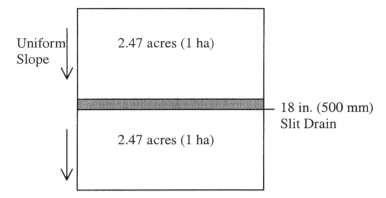

Figure 11.5. Shows an 18 in. (500 mm) wide slit drain installed in between two football fields, so as to collect all of the surface run-off from the top field before it reaches the bottom field. The profile in the trench is the same as in Figure 11.3.

The sand in the trench must be a uniform depth of 10 in. (250 mm) and the extra depth of the trench is taken up with gravel. This perches a water table and allows the grass to grow and not to be droughty in the summertime. It is essential to ensure that the sod is thoroughly washed to remove any fines from it and to allow the drain to accept water at a rate of at least 15 in. (400 mm)/hr.

Such a drain will accept water flowing off 2.5 ac (1 ha) when it is being generated by continuous rainfall of about 0.2 in. (5 mm)/hr. This is quite heavy rain, considering that the first rain will have to wet the profile and saturate the soil surface before it begins to run off. This strategy has worked very well in Canberra—and the bottom area, i.e., the area downslope, only receives the water that falls on it, rather than the huge volumes of water that used to reach it before. The playability of these lower areas has improved dramatically, as have their maintenance costs.

Golf Courses

The principle of slit drainage can be used on golf courses in a wide range of situations. One is where fairways or sections of fairway have been constructed without surface slope deflecting water to either side. Consequently, water flows straight down the length of the fairway.

The principle can be used around a green to collect water that runs off the green, which often gets more water than the surrounds, making them soft and boggy. It is important to ensure that the sand is a uniform 10 in. (250 mm) deep and extends right up to the surface and that the sod that covers it is washed free of all soil.

These slit drains can be installed around sumps in fairways as an insurance policy against their blocking. Grass clippings, leaves, paper, and plastic can quickly block grated sumps during heavy rain. If this water stays in these depressions for an extended period, the soil surrounding the low point around the sump becomes saturated. Once saturated it is vulnerable to damage.

These drains should be installed about 18 in. to 3 ft (500 mm to 1 m) from the sump. The trench can be circular using flexible corrugated drainage pipe. The pipe is then simply fed into the sump. Once again it is imperative that the sand extends right to the surface and that the sod is washed free of soil.

On fairways there is no need to have the slits at 6.6 ft (2 m) spacings as outlined above; however, if they are further apart the trenches have to be wider and the pipes in the bottom of the trenches have to be larger to cope with the larger volumes of water.

Racecourses

Slit drainage can be used to drain flatter areas on racecourses, but there are some problems. If the slits are installed into courses grassed with Bermuda (couch) or kikuyu, the rapid covering over of damage to the top of the trench usually helps keep the top of the trench "open" and it will continue to receive water.

However, on courses that use cool-season grasses, once a divot is taken out of the top of the trench it is very difficult to reestablish grass into the sand in the trench. Once the top of the trench is bare, it tends to allow the edges of the trench to be damaged. The topsoil from the surrounding area contaminates the trench and reduces the hydraulic conductivity. One meeting in very wet conditions can render large sections of the drains useless.

Once the slit drains are installed all topdressing and divot replacement material must be of the same material as the sand in the slits.

Decompaction activities can also damage the integrity of the slits and cause contamination into the trench from the adjacent topsoil.

On most racecources slit drainage has limited value and life.

Note

When quantities are converted into metric, the answers are not always the same as for imperial, as this is due to rounding off errors; e.g., 2 m is 6.562 ft, which is rounded up to 6.6 ft.

CHAPTER **12**

Methodology for Determining Soil Hydraulic Conductivity

Introduction

In this book we have referred to the ***drop test*** in several situations as a measure of saturated hydraulic conductivity. There are many different methods used for the measurement of hydraulic conductivity in soils; these different methods can give different results.

Over the past eight years Dr. Jakobsen has been developing a very simple and effective test for saturated hydraulic conductivity that can be used in the meaningful assessment of soils and sands for sportsturf and horticulture. This method requires no expensive laboratory equipment, and has proved to be highly repeatable.

Many of the traditional methods for determining hydraulic conductivity involve dropping a hammer or some sort of weight onto the soil sample, which is held in a steel container. There is one major problem with this method of applying compaction, and that is the friction loss of the impact with the soil that is very close to the edge of the container.

This causes a deceleration of the energy being applied close to the edge. This means that the soil close to the edge of the container is not compacted to the same density as the soil near the centre of the container. When water flows down this column, it passes through the less dense edges of the soil in the container more quickly than through the adjacent soil.

There is also a variation in the compaction, or bulk density of the sample, from the top to the bottom. This also means that the compaction needed by the hammer or dropping weight method has to be high to achieve good compaction of the whole sample.

If the whole container, including the soil inside it (Handreck and Black, 1984), is dropped onto a flat hard surface, then much less energy has to be applied to achieve the same compaction of the sample. The soil and the edge of the container decelerate at the same rate, so there is a minimum of edge effect in this method, and the soil compacts more uniformly through the whole sample than from the drop hammer method.

The apparatus used for this procedure is ordinary laboratory equipment, including a drying oven, suction plate, vernier calipers, distilled water, and a balance. There is no complex machine required to apply compaction, and there is no special apparatus required to keep a constant head of water for special stainless steel tubes.

A 4 L ice cream container, or large beaker, is ideal to place the tubes of soil in for wetting up from the bottom.

PVC water pipe (25-mm-diameter) can be used for the tubes. Cut into 150-mm lengths, ensuring that the ends are cut off exactly at right angles and the edges smoothed so they are even. Each tube has a piece of gauze glued to the bottom to prevent soil from falling out, while still allowing water to freely flow through it.

Note

It is very important to ensure that the sample is at field capacity when the soil is placed in the tubes prior to the test. Failure to ensure this will bring about the wrong result. It is also very important to ensure that the tube is dropped from the correct height every time and that it falls vertically. If the tube falls on an angle rather than flat on the bottom, it can decompact the sample.

With practice and care, the method is very reliable and staff can be trained to carry it out with a high degree of repeatability.

The methods have been written in the format of an Australian Standard and are as follows:

1. Method for Determining Soil Water Content

General

Before any soil is used for testing, the whole sample received must be made homogeneous.

Pretreatment of soil samples before physical testing can affect the results, in particular for hydraulic conductivity, but also to a lesser degree for compaction and water holding capacity. Thus, the screening and mixing of samples, as needed to obtain representative subsamples for different tests, must be carried out with a minimum impact on the soil. Otherwise the pretreatment may easily cause effects on the soil that will not occur in the field operations.

Many soils are destabilised and slake easily after being worked while wet; after drying, the soil may regain its former stability. If such a soil is stored moist in a stockpile to be used for an irrigated turf area, then drying of the soil is unlikely to occur in the field, and drying should be avoided in the laboratory prior to testing.

Dry and lumpy clay soil should be moistened first, so that it crumbles easily and can be passed through the screen with a minimum of effort. Crushing dry lumps with a hammer will produce a lot of dust, which is not characteristic of the soil in nature.

Dry soil or gravel is difficult to sample representatively, because fine particles fall to the bottom between larger ones. Moistening the sample to a water content just below its field capacity will make all dust cling to larger particles. Such a sample can be mixed well and representative subsamples taken out. If the water content is below the lower plastic limit, mixing will not destabilise the soil and cause it to slake.

If the soil is dry, then add water to what is judged to be near field capacity, e.g., for a gravel 1 to 1.5 %, a sand ~ 5 to 8 %, a pug soil ~ 20 to 30 %. Give clay soil time to absorb the water so that no muddy lumps are formed during screening and mixing.

Scope

This standard follows the procedures outlined in Australian Standard 1289.1 and 1289.2.1.1.

Apparatus

    a. A drying oven
    b. A heat-resistant and corrosion-resistant container
    c. Analytical balance
    d. 12-mm screen
    e. Desiccator containing anhydrous silica gel

Procedure

Crush any big lumps by hand so that the soil can be passed through a 12-mm screen.

Mix the sample well and store in a plastic storage container (a) with an airtight lid.

1. Weigh a clean and dry container (b) in grams, to 2 decimal places, **T**.
2. Take a representative subsample of soil from the storage container (a) and place in the drying container (b). Weigh container (b) with wet soil, **TSW**.
3. Place sample in drying oven at 105°C for at least 12 hours, or until its weight becomes constant.
4. Allow sample to cool in a desiccator for 10–20 minutes and weigh again, **TS**.

Calculations

Calculate the gravimetric water content from:

$$W\% = \frac{TSW - TS}{TS - T} \qquad (1)$$

The relation between the gravimetric water content and the volumetric water content of a soil at a given bulk density is:

$$\text{water vol.\%} = w\% \times \text{bulk density (g/cm}^3) \qquad (2)$$

For tests made on wet soil, but where the amount of dry soil must be known, e.g., hydraulic conductivity, a conversion factor, **C**, may be useful:

$$C = \frac{TS - T}{TSW - T} \tag{3}$$

Now use:

$$\text{Dry soil, g} = C \times \text{Wet soil, g} \tag{4}$$

If the moist soil is kept in a closed plastic container, then the measured water content and the value of C may be valid for a week or more.

2. Procedure for Preparing Samples to be at Field Capacity for Testing (Canberra Landscape Guidelines, 1993)

Scope

This standard sets out a method for the laboratory determination of the field capacity of a soil.

Apparatus

a. Ceramic suction plate, adjusted to 1 m suction (–10 kPa)*
b. Plastic rings, approximately 30 mm high and 50 mm in diameter
c. Heat-resistant and corrosion-resistant containers
d. Vernier calipers
e. Analytical balance
f. Drying oven complying with AS 1289.0
g. Wash bottle

---

*Note: Flexible tubing is used to join the suction plate to a water reservoir. The porous plate is saturated with water, and air is removed from the cavity behind the plate and from the flexible tubing. By adjusting the free water surface in the reservoir to 1 m below the upper plate surface, 1 m suction is achieved. This system is usually referred to as a hanging water column.

Procedure

1. Wet the suction plate and make sure there are no air bubbles in the drain tubes.
2. Place the empty plastic rings on the suction plate and fill the rings with soil. Pack the soil samples to ensure good contact with the suction plate.
   Wet the samples thoroughly using a wash bottle.
   Cover the suction plate and leave the samples to drain for at least 16 hours.
3. Weigh the empty drying containers, **T**g.
   Using vernier calipers, measure the depth of soil, **H**mm.
   Transfer the samples from the rings to the heat-resistant containers and record the wet weight, **TSW**g.
   Record the dry weight, **TS**g after drying at 105°C for at least 12 hours.

Calculations

Soil bulk density (D_b):

$$D_b = \frac{TS - T}{H \times 2.02} \ g/cm^3 \qquad (1)$$

Gravimetric water content (w):

$$W = \frac{TSW - TS \times 100}{TS - T} \% \qquad (2)$$

This description is taken from Loveday, 1974:

3. Procedure for the Determination of Saturated Hydraulic Conductivity in Soils and Sands

General

The saturated hydraulic conductivity, **K**, of a soil refers to the movement of water through the soil profile when it is saturated (completely filled with water). It is the coefficient, **K**, in Darcy's equation:

$$\text{Rate of flow} = K\frac{dh}{dx} \text{ mm/hr} \qquad (1)$$

where (**dh/dx**) is the driving force, i.e., change in water pressure with distance.

In the following test **K** (saturated) is measured under conditions of a falling head of water, **H**. At the time when the surface of the ponded water reaches the soil surface, it falls at a rate equal to **K**.

The samples used for testing are compacted to a range of soil densities, which are expected to cover the range occurring in the field after several years of use. A very light level of compaction is applied as well, because some soils will slump during wetting and then become very slow-draining.

Scope

This standard sets out a method for laboratory determination of the hydraulic conductivity of a soil.

Apparatus

The following apparatus is required:

a. A drying oven
b. 6 Plastic tubes, 150 mm long, 30-mm in diameter, with nylon gauze fitted on one end. The gauze should allow free drainage of water, but not of soil
c. Plunger, comprised of a plastic tube with a rubber stopper on one end and fitting neatly inside the plastic tubes
d. Analytical balance
e. Vernier calipers
f. Large plastic container, deeper than 150 mm to fit the 6 plastic tubes
g. Free-draining surface
h. Heat-resistant container
i. Hard flat surface, e.g., steel base of retort stand

j. Stopwatch

k. Wash bottle

Procedure

Sample preparation

1. The soil should have a water content near field capacity prior to testing, i.e., the water content held against a suction of 10 kPa or 1 m hanging water column.

2. Mix the moist soil sample well to make sure the moisture content is uniform and that no wet lumps exist. Keep the soil in a closed container to avoid water loss during the test. Take out a sample for determination of the exact water content. (See Method 1.)

3. Prior to filling each tube with the moist soil, place the tube on the balance and tare the balance.

4. Fill the tube with soil. While filling, stand the tube on a flat surface to ensure that the soil does not bulge out at the bottom.
   Compact the soil in the first tube by dropping it once from a height of 150 mm onto a hard, flat surface. Ensure that the tube is kept upright and is not allowed to fall over.
   Lightly firm the soil surface down to the same level of compaction as the rest of the soil in the tube using a plunger. Do not use excessive force.

5. Weigh the tube immediately after filling and compaction to prevent weight loss due to evaporation. Record the weight of wet soil in the tube, **SW**.
   Using vernier calipers, measure and record the distance from the rim of the tube to the compacted soil surface, **h**.

6. Repeat steps 4 to 6 on the remaining tubes, applying increasing levels of soil compaction for each tube by doubling the number of times, **N**, the tubes are dropped, i.e., 2, 4, 8, 16, and 32.
   Record the values of **SW, h,** and **N** for each sample.

7. Place the tubes in the plastic container, and slowly fill this container with water until the level approaches the outside rim of the tubes. Allow the water to rise up through the soil to displace most of the soil air.

Only fill the tubes from the top after the water surface inside them is close to the rim, or at least after the soil surface is under water. This should be done carefully with a wash bottle so as to minimise disturbance of the soil surface.

Testing procedure

1. Lift the tubes out of the plastic container and place them on a free-draining surface.
   For fast-draining sands lift one tube at a time and use a stopwatch to record the time, **t**, for the water to fall from the rim of the tube to the soil surface. For a coarse sand this time may be as little as 5 seconds.
   For slow-draining samples, when **t** is likely to exceed 10–20 minutes, it becomes difficult and too time-consuming to get the exact time when water disappears from the soil surface. In these instances lift all the samples out and record their starting time. Measure the distance from the rim to the water surface at intervals (e.g., 10, 20, and 30 minutes; these intervals should be varied according to the rate of drainage): **h1, t1; h2, t2; h3, t3;** etc.
   Take three readings or more of each sample.
2. When each sample is fully drained, measure with vernier calipers the distance from the rim to the soil surface, **h**, again, as the soil may have slumped after wetting.

Calculations

The conductivity, **K,** and the bulk density, **D$_b$,** for the soil of each tube is calculated by use of equations (2), (3), and (4).

The measurements were made under a decreasing head of water, **H1** = 150 – **h1**, **H2** = 150 – **h2**, **H3** = 150 – **h3**, where **h1, h3** is the distance from the rim to the water surface and 150 is the tube height, all in mm.

The decreasing head causes a decreasing rate of flow, which is compensated for in the equation for calculation of the hydraulic conductivity, **K**:

$$K = \frac{1}{t} \times \ln\left[H_1 / H_2\right] \text{ mm/hr} \qquad (2)$$

where l = 150 – h, height of soil column in mm; and t is the time in hours between the measurements **H1** and **H2**. Repeat the calculation with **H2** and **H3** and the corresponding value of t.

The weight of dry soil, **S**, in a tube is calculated by:

$$S = SW \div \frac{1.0 + W\,\%}{100}\ g \tag{3}$$

The bulk density is:

$$D_b = \frac{S}{0.1 \times 1 \times A} \tag{4}$$

where l is the height of soil in mm (150 – h), and **A** is the cross-sectional area of the tube (7.07 cm^2 for a 30-mm-diameter tube).

If no additional compaction test is made, then the sample compacted by 16 drops of 150 mm is used as a standard level of compaction for irrigated turf, and 8 drops for shrub beds and nontraffic areas.

All values of **K** for tubes of compaction by 16 drops and less should be more than 5 mm/hr.

USGA Recommendations for a Method of Putting Green Construction

by the USGA Green Section Staff

Following is the 1993 revision of the USGA Recommendations for a Method of Putting Green Construction.

Step 1. The Subgrade
Step 2. Drainage
Step 3. Gravel and Intermediate Layers
Step 4. The Root Zone Mixture
Step 5. Top Mix covering, Placement, Smoothing, and Firming
Step 6. Seed Bed Preparation
Step 7. Fertilization

Step 1. The Subgrade

The slope of the subgrade should conform to the general slope of the finished grade. The subgrade should be established approximately 16 inches (400 mm) below the proposed surface grade—18 to 20 inches (450 to 500 mm) when an intermediate layer is necessary—and should be thoroughly compacted to prevent further settling. Water collecting depressions should be avoided.

If the subsoil is unstable, such as with an expanding clay, sand, or muck soil, geotextile fabrics may be used as a barrier between the subsoil and the gravel blanket. Install the fabric as outlined in Step 2.

Construct collar areas around the green to the same standards as the putting surface itself.

Step 2. Drainage

A subsurface drainage system is required in USGA greens. A pattern of drainage pipes should be designed so that the main line(s), with a minimum diameter of 4 inches (100 mm), is placed along the line of maximum fall. Four-inch (100 mm) diameter laterals shall run up and across the slope of the subgrade, allowing a natural fall to the main line. Lateral lines shall be spaced not more than 15 feet (5 m) apart and extended to the perimeter of the green. Lateral lines should be placed in water-collecting depressions, should they exist. At the low end of the gradient, adjacent to the main line's exit from the green, drainage pipe should be placed along the perimeter of the green, extending to the ends of the first set of laterals. This will facilitate drainage of water that may accumulate at the low end of that drainage area. Drainage design considerations should be given to disposal of drainage waters away from play areas, and to the laws regulating drainage water disposal. PVC or corrugated plastic drainage pipe is preferred. Where such pipe is unavailable, clay or concrete tile is acceptable. Waffle drains or any tubing encased in a geotextile sleeve are not recommended. Drainage trenches 6 inches (150 mm) wide and a minimum of 8 inches (200 mm) deep shall be cut into a thoroughly compacted subgrade so that drainage lines slope uniformly. Spoil from the trenches should be removed from the subgrade cavity, and the floor of the trench should be smooth and clean. If a geotextile fabric is to be used as a barrier between an unstable subsoil and the gravel drainage blanket, it should be installed at this time. Under no circumstances should the fabric cover the drainage pipes or trenches. A layer of gravel (see Step 3 for size recommendations) should be placed in the trench to a minimum depth of 1 inch (25 mm). It may be deeper, as necessary, to ensure a positive slope along the entire run of drainage lines. If cost is a consideration, gravel sized 1/4 to 1 inch (6 to 25 mm) may be used for the drainage trench only. All drainage pipe should be placed on the gravel bed in the trench, assuring a minimum positive slope of 0.5 percent. PVC drain pipe, if used, should be placed in the trench with the holes facing down. Backfill with additional gravel, taking care not to displace any of the drainage pipe.

Step 3. Gravel and Intermediate Layers

Place grade stakes at frequent intervals over the subgrade and mark them for the gravel drainage blanket layer, intermediate layer (if included), and root zone layer.

The entire subgrade then shall be covered with a layer of clean, washed, crushed stone or pea gravel to a minimum thickness of four inches (100 mm), conforming to the proposed final surface grade to a tolerance of ± 1 inch.

Soft limestones, sandstones, or shales are not acceptable. Questionable materials should be tested for weathering stability using the sulfate soundness test (ASTM C-88). A loss of material greater than a 12 % by weight is unacceptable.

The LA Abrasion test (ASTM C-131) should be performed on any materials suspected of having insufficient mechanical stability to withstand ordinary construction traffic. The value obtained using this procedure should not exceed 40. Soil engineering laboratories can provide this information.

The need for an intermediate layer is based on the particle size distribution of the root zone mix relative to that of the gravel. When properly sized gravel (see Table 1) is available, the intermediate layer is not necessary. *If the properly sized gravel cannot be found, an intermediate layer must be used.*

Table 1. Particle size description of gravel and intermediate layer materials.

Material	Description
Gravel: Intermediate layer is used	Not more than 10 % of the particles greater than 1/2" (12mm)
	At least 65 % of the particles between 1/4" (6mm) and 3/8" (9mm)
	Not more than 10 % of the particles less than 2 mm
Intermediate Layer Material	At least 90 % of the particles between 1 mm and 4 mm

Table 2. Size recommendations for gravel when intermediate layer is not used.

Performance Factors	Recommendation
Bridging Factor	D15 (gravel) less than or equal to 5 X D85 (root zone)
Permeability Factor	D15 (gravel) greater than or equal to 5 X D15 (root zone)
Uniformity Factors	D90 (gravel) / D15 (gravel) is less than or equal to 2.5
	No particles greater than 12 mm
	Not more than 10% less than 2 mm
	Not more than 5% less than 1 mm

A. Selection and Placement of Materials When the Intermediate Layer Is Used

Table 1 describes the particle size requirements of the gravel and the intermediate layer material when the intermediate layer is required.

The intermediate layer shall be spread to a uniform thickness of two to four inches (50 to 100 mm) over the gravel drainage blanket (e.g., if a 3-inch depth is selected, the material shall be kept at that depth across the entire area), and the surface shall conform to the contours of the proposed finished grade.

B. Selection of Gravel When the Intermediate Layer Is Not Used

If an appropriate gravel can be identified (see Table 2), the intermediate layer need not be included in the construction of the green. In some instances, this can save a considerable amount of time and money.

Selection of this gravel is based on the particle size distribution of the root zone material. The architect and/or construction superintendent must work closely with the soil testing laboratory in selecting the appropriate gravel. Either of the following two methods may be used:

1. Send samples of different gravel materials to the lab when submitting samples of components for the root zone mix. As a general guideline, look for gravel in the 2 mm to 6 mm range. The lab first will determine the best root zone mix,

and then will test the gravel samples to determine if any meet the guidelines outlined below.

2. Submit samples of the components for the root zone mix, and ask the laboratory to provide a description, based on the root zone mix tests, of the particle size distribution required of the gravel. Use the description to locate one or more appropriate gravel materials, and submit them to the laboratory for confirmation.

Gravel meeting the criteria below will not require the intermediate layer. It is not necessary to understand the details of these recommendations; the key is to work closely with the soil testing laboratory in selecting the gravel. Strict adherence to these criteria is imperative; failure to follow these guidelines could result in greens failure.

The criteria are based on engineering principles which rely on the largest 15% of the root zone particles "bridging" with the smallest 15% of the gravel particles. Smaller voids are produced, and they prevent migration of root zone particles into the gravel yet maintain adequate permeability. The D85 (root zone) is defined as the particle diameter below which 85% of the soil particles (by weight) are smaller. The D15 (gravel) is defined as the particle diameter below which 15% of the gravel particles (by weight) are smaller.

- For bridging to occur, the D15 (gravel) must be less than or equal to five times the D85 (root zone).
- To maintain adequate permeability across the root zone/gravel interface, the D15 (gravel) shall be greater than or equal to five times the D15 (root zone).
- The gravel shall have uniformity coefficient (Gravel D90/Gravel D15) of less than or equal to 2.5.

Furthermore, any gravel selected shall have 100% passing a 1/2" (12 mm) sieve and not more than 10% passing a No. 10 (2 mm) sieve, including not more than 5% passing a No. 18 (1 mm) sieve.

Step 4. The Root Zone Mixture

Sand Selection: The sand used in a USGA root zone mix shall be selected so that the particle size distribution of the *final root zone mixture* is as described in Table 3.

Soil Selection: If soil is used in the root zone mix, it shall have a minimum sand content of 60%, and a clay content of 5% to 20%. The final particle size distribution of the sand/soil/peat mix shall conform to that outlined in these recommendations, and meet the physical properties described herein.

Organic Matter Selection: Although the USGA encourages the use of organic matter in root zone mixtures due to its beneficial properties, it is recognized that some sands may meet the particle size and physical properties guidelines without modification. Therefore, the guidelines no longer specify a minimum organic matter percentage. Note: Since such sands rarely occur, the vast majority of sands must be modified with organic matter to meet the required physical characteristics.

Peats—The most commonly used organic component is a peat. If selected, it shall have a minimum organic matter content of 85% by weight as determined by loss on ignition (ASTM D 2974-87 Method D).

Other organic sources—Organic sources such as rice hulls, finely ground bark, sawdust, or other organic waste products are acceptable if composted through a thermophilic stage, to a mesophilic stabilization phase, and with the approval of the soil physical testing laboratory. Composts shall be aged for at least one year. Furthermore, the root zone mix with compost as the organic amendment must meet the physical properties as defined in these recommendations.

Composts can vary not only with source, but also from batch to batch within a source. Extreme caution must be exercised when selecting a compost material. Unproven composts must be shown to be nonphytotoxic using a bentgrass or bermudagrass bioassay on the compost extract.

Inorganic and Other Amendments: Inorganic amendments (other than sand), polyacrylamides, and reinforcement materials are not recommended at this time in USGA root zone mixes.

Physical Properties of the Root Zone Mix: The root zone mix shall have the properties summarized in Table 4, as tested by USGA protocol (proposed ASTM Standards).

Under the heading Saturated Conductivity in Table 4, Normal range refers to circumstances where normal conditions prevail for growing the desired turfgrass species. Accelerated range refers to conditions where water quality is poor, cool season turfgrass species are being grown out of range of adaptation, or dust storms or high rainfall events are common.

Table 3. Particle size distribution of USGA root zone mix.

	Particle	Recommendation
Name	**Diameter**	**(by weight)**
Fine Gravel	2.0–3.4 mm	Not more than 10% of the
Very coarse sand	1.0–2.0 mm	total particles in this range, including a maximum of 3% fine gravel (preferably none)
Coarse sand	0.5–1.0 mm	Minimum of 60% of the particles must fall in this range
Medium sand	0.25–0.50 mm	Not more than 20% of the
Fine sand	0.15–0.25 mm	particles may fall within this range
Very fine sand	0.05–0.15 mm	Not more than 5%
Silt	0.002–0.05 mm	Not more than 5%
Clay	less than 0.002 mm	Not more than 3%
Total fines	Very fine sand + silt + clay	Less than or equal to 10%

Table 4. Physical properties of the root zone mix.

Physical Property	Recommended Range
Total porosity	35%–55%
Air-filled porosity	15%–30%
Capillary porosity	15%–25%
Saturated conductivity Normal range:	6–12 inches/hr (15-30 cm/hr)
Accelerated range:	12–24 inches/hr (30-60 cm/hr)

Related Concerns

It is absolutely essential to mix all root zone components off-site. No valid justification can be made for on-site mixing, since a homogeneous mixture is essential to success.

A quality control program during construction is strongly recommended. Arrangements should be made with a competent laboratory to routinely check gravel and/or root zone samples brought to the construction site. It is imperative that these materials conform to the recommendations approved by the laboratory in all respects. Some tests can be performed on site with the proper equipment, including sand particle size distribution.

Care should be taken to avoid overshredding the peat, since it may influence performance of the mix in the field. Peat should be moist during the mixing stage to ensure uniform mixing and to minimize peat and sand separation.

Fertilizer should be blended into the root zone mix. Lime, phosphorus, and potassium should be added based on a soil test recommendation. In lieu of a soil test, mix about 1/2 pound of 0-20-10 or an equivalent fertilizer per cubic yard of mix.

Step 5. Top Mix Covering, Placement, Smoothing, and Firming

The thoroughly mixed root zone material shall be placed on the green site and firmed to a uniform depth of 12 inches (300 mm), with a tolerance of ± 1/2 inch. Be sure that the mix is moist when spread to discourage migration into the gravel and to assist in firming.

Step 6. Seed Bed Preparation

Sterilization: Sterilization of the root zone mix by fumigation should be decided on a case by case basis, depending on regional factors. Fumigation always should be performed:

1. In areas prone to severe nematode problems.
2. In areas with severe weedy grass or nutsedge problems.
3. When root zone mixes contain unsterilized soil.

Check with your regional office of the USGA Green Section for more information and advice specific to your area.

Step 7. Fertilization

Contact your regional USGA Green Section office for establishment fertilizer recommendations and grow-in procedures.

Bibliography

Adams, W.A. and R.J. Gibbs. 1994. *Natural Turf for Sport and Amenity.* CAB International, Wallingford UK, 1st ed. 404 pp.

Department of Urban Services, ACT Government. 1993. *Canberra Landscape Guidelines, Soil Testing Procedure LG B22.* Canberra, Aust. 456 pp.

Handreck, K. and N. Black. 1984. *Growing Media for Ornamental Plants and Turf.* New South Wales University Press, Kensington, Aust. 401 pp.

Loveday, J. 1974. *Methods for Analysis of Irrigated Soils.* Technical Communication No. 54. Commonwealth Agricultural Bureaux, Buckinghamshire, England. 171 pp.

Marshall, T. J., J.W. Holmes, and C.W. Rose. 1996. *Soil Physics.* Cambridge University Press, New York, 3rd ed., 453 pp.

McIntyre, D.K. and S. Hughes. 1988. *Turf Irrigation, Management Aid No 2.* Royal Aust. Inst. of Parks and Recreation, Canberra, Aust. 53 pp.

McIntyre, D.K. and B. Jakobsen. 1998. *Drainage for Sportsturf and Horticulture.* Horticultural Engineering Consultancy, Canberra, Aust. 162 pp.

Russell, E.W. 1962. *Soil Conditions and Plant Growth.* Longmans, London, UK. 9th ed. 688 pp.

Standards Australia. *Methods of Testing Soils for Engineering Purposes, Australian Standard 1289.–Method 1289.1., 1991. Preparation of disturbed soil samples for testing;* ISBN No. 0 7262 6924 7.

Standards Australia. *Methods of Testing Soils for Engineering Purposes, Australian Standard 1289.–Method 1289.2.1.1., 1992. Soil moisture tests— determination of the moisture content of a soil;* ISBN No. 0 7262 7389 9.

Standards Australia. *Methods of Testing Soils for Engineering Purposes, Australian Standard 1289.–Method 1289.3.8.1., 1997. Soil classification tests—determination of Emerson Class number of a soil;* ISBN No. 0 7337 1089 1.

United States Golf Association. 1993. *USGA Recommendations for a Method of Putting Green Construction.* USGA Green Section Record. March/April 1993.

Yong, R.N. and B.P. Warkentin. 1975. *Soil Properties and Behaviour.* Elsevier Scientific Publishing Co., Amsterdam, Neth., 1st ed. 449 pp.

Index

Soil structure   1–5
Soil texture, effect on infiltration   17
Springs or wet spots   106
Sub-soil drains
    designing systems   103–123
    how to calculate spacings
        89–102
    how water enters   57–63
    in flower and shrub beds
        114–121
    in racecourses   121–123
    installing   125–140
    wrong assumptions   42
Suction   15, 21–24, 59, 71, 73, 76,
    87
    nil (or zero) suction   51, 58, 60
    of gravel in perched water tables
        84
    of one metre, soil test   90, 122,
        183
Surface tension   8–15, 23, 25, 60,
    62, 75, 76, 82, 83

T

Traffic
    effect on drainage   27, 30, 31,
        51, 119, 147, 163
Trapped air bubbles   16, 21

U

Uncoiling pipes   133, 137
Uneven irrigation   49
Unstable soil   3
USGA sand   11, 13, 34, 58, 74,
    76–78, 80, 82, 84, 90, 97,
    127, 129, 130, 164, 170, 171,
    174

V

Velocity of water in pipes   108,
    135
Very fine sand   2, 38
Voids   4, 10, 31, 36, 37, 125–130

W

Water holding capacity   5, 10, 15,
    21, 36, 37, 90, 121, 181
Waterlogged conditions   24, 32, 96,
    102, 115, 116, 144, 152
Water retention curve. *See* Moisture
    release curve
Water storage in soils   22
Wet spots   106
Wetting front   16
Wilting point   13, 20, 21, 23, 24,
    59